Destiny Ma

Making Star Trek Real: The Metric Engineering of Warp Drive and Star Gate Time Travel

The World According to Jack

The Life and Times of a Theatrical Physicist

The Creation of Adam, Michelangelo, Sistine Chapel, The Vatican

A pageant, a post-modern "Pilgrim's Progress" and "Divine Comedy", a rich tightly woven tapestry of uncanny synchronicities; a true larger than life panoramic epic; an anthology of North Beach San Francisco writers created in Super Vision

By
Jack Sarfatti

The man who made "The Dancing Wu Li Masters" possible and co-author of the original
"Space-Time and Beyond"
"A Harry Potter for enlightened adults." Frank Lauria, author "Raga Six"

ISBN: 0-7596-9688-8 (e-book)
ISBN: 0-7596-9689-6 (Paperback)
ISBN: 0-7596-9690-X (Dustjacket)
ISBN: 0-7596-9691-8 (Rocket Book)

Library of Congress Control Number: 2002100577

This book is printed on acid free paper.

Printed in the United States of America
Bloomington, IN

1stBooks - rev. 10/8/02

"I believe in destiny. Destiny commands and I obey."
Allegedly spoken by Winston Churchill according to HBO's "The Gathering Storm"
http://www.hbo/films/gatheringstorm/

Including key contributions from Stephen Schwartz, Saul Paul Sirag, Jagdish Mann, Kim Burrafato, Creon Levit and Hyman Sarfatti at Age 88 with cartoons by Norman Quebedeau and James Anderson

Collage by Allen Cohen.

Download the book within the book 40 cartoons & pictures
http://qedcorp.com/destiny/STB.doc

Table of Contents

vi

"It's tricky to speculate openly about time travel. One risks either an outcry at the waste of public money...or a demand that the research be classified for military purposes....There are only a few of us foolhardy enough to work on a subject that is so politically incorrect in physics circles. We disguise the fact by using technical terms that are code for time travel (p.133)...we have no reliable evidence of visitors from the future. I'm discounting the conspiracy theory that UFOs are from the future and that the government knows and is covering up. Its record of cover-ups is not that good (p.142).... You might wonder if this chapter is part of a government cover-up on time travel. You might be right. (p.153)"
Stephen Hawking "The Universe in a Nutshell" (Bantam Books, 2002)

"What I cannot create. I do not understand." Richard Feynman

"To advance scientific knowledge, pick a man of genius, give him money, and let him alone." James B. Conant (President of Harvard), cited in "Tuxedo Park" by Jennet Conant

http://stardrive.org/cartoon/MagicBean.html

http://stardrive.org/cartoon/spectra.html

http://stardrive.org/cartoon/coffee.html

http://stardrive.org/title.shtml

"For no man can write anything who does not think that what he writes is for the time the history of the world; or do anything well who does not esteem his work to be of importance... there is throughout nature something mocking, something that leads us on and on, but arrives nowhere; All promise outruns the performance. We live in a system of approximations."
Ralph Waldo Emerson, Essay on Nature (1844)

Honey: "What's this business with the forehead?"
Fred: "Mental telepathy."
Honey: "I can tell what they're thinking about from here."
    "Flying Down to Rio" Fred Astaire & Ginger Rogers (1933)

"F.SCOTT FITZGERALD: The rich are different from us.
ERNEST HEMINGWAY: Yes, they have more money."
"More Is Different", P.W. Anderson, SCIENCE, 177, pp.393-396, 4 August, 1972

"But there's more to it than that. Wolfram is selling not just a theory, but a story – the story of a genius who retreats from the misunderstanding and jealousy of smaller minds. Some call the genius crazy, but he doesn't mind. Some raise technical objections to his theory. But the true genius knows that true ideas are always simple and that technicalities are just the so-called experts' way of hiding the inadequacy of their ideas. He should read and give credit to the work of his peers? What they know, he doesn't need. Anyway, reading would waste time: time he needs to erect the intellectual monument history will thank him for. Sound familiar? You've heard it a lot lately. It's the story of *A Beautiful Mind*: Ron Howard's cartoon version of John Nash is shown scribbling in the library a lot, but does his real work while helping his less inspired classmates pick up girls in bars." Jordan Ellenberg, "Blinded by Science" MSN, July 2, 2002. Thanks to Marcello Truzzi.

SUNDAY INTERVIEW - The Universe, As Seen From North Beach by Stephen Schwartz, San Francisco Chronicle, August 17, 1997

"With infectious enthusiasm for his subject, Jack Sarfatti explains how physics has replaced philosophy as an over-arching discipline that spans the once discontinuous worlds of science and the humanities. Since he first came to the Bay Area in the mid-1970s, physicist Jack Sarfatti has been a provocative presence in local intellectual life.

Leaping from North Beach cafes to leading policy think-tanks, he has cut a broad intellectual swath, challenging the preconceptions of poets, political thinkers and physicists alike.

With a background in quantum theory, he claims to break new ground in scientific understanding of the eternal questions: 'Who are we? Where do we come from? Where are we going?'

In this interview, he discusses the breakdown of a paradigm that, for centuries in the West, has viewed science and humanistic thought as irrevocably separated."

Breakdown of a Paradigm

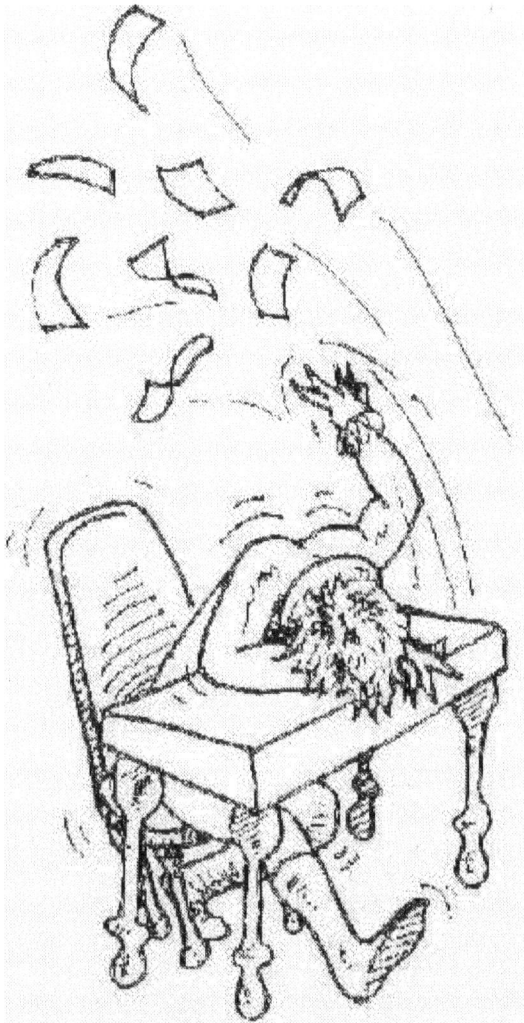

THE BREAKDOWN
OF CLASSICAL PHYSICS

Cartoon by James Anderson

"Science proceeds as if the past were the home of explanation; whereas the future, and the future alone, holds the key to the mysteries of the present." Henry Dwight Sedgwick, An Apology for Old Maids, House of Sorrow (1908)

"In the half-century that North Beach has supplied a haven for waves of brilliant and erratic thinkers, few have rivaled Jack Sarfatti for comprehensive weirdness and what just might be original genius."

Another Eccentric Genius in North Beach? By Jerry Carroll, San Francisco Chronicle, May 11,1981, p. 23

"Creation is emerging as even stranger than we thought."[1]

Sir Martin Rees, Cambridge, 9/9/2001, London Sunday Times

Indeed it is even stranger than possibly, my old acquaintance, Martin is willing to think!

This book is, by far, the weirdest true story ever told. It tells how our Super Cosmos, of parallel universes floating in hyperspace, is connected together by Star Gate time travel machines carrying "The Masters of Hyperspace"[2] back to us from our futures. We all come into being and becoming from the post-quantum Mind of God. Participate in the real life thrilling Psi Wars adventure, the amazing paranormal flying saucer Grail Quest of visionary physicist Jack Sarfatti

---

[1] Light *appears* to have moved faster in our early universe which is now speeding up rather than slowing down. I explain this qualitatively in terms of a variable local zero point energy Λ "quintessent" field in the physical vacuum that is a *zero entropy* coherent "superfluid" of bound virtual electron-positron pairs in a "Bose-Einstein condensate". Einstein's gravity of curved spacetime comes from smooth variations in the phase of the vacuum superfluid. Quintessence comes from smooth variations in the amplitude of the vacuum superfluid. The Einstein vacuum normally does not *gravitate*, however, under certain conditions finite regions of the vacuum can either gravitate forming "dark matter", or *anti-gravitate* forming "dark energy". This neatly explains, in one bold stroke, both the missing mass of the universe and why it is accelerating in addition to other things like star gate time machines and weightless warp drives for alleged flying saucers and last, but not least, why time *appears* to flow from past to future this update note was added on August 11,2002. One can see below the progress, the continual refining of this *surprising* unexpected sudden original idea of mine, perhaps *from* "remote viewing" the future, from its earlier vague conception in 2001 when most of this book was written.

[2] Michio Kaku's book "Hyperspace" and Stephen Hawking's "The Universe in a Nutshell".

4

and "the others" as they decode The Cabala's[3] "Cipher of Genesis" demystifying the enigma of consciousness, a body guard of spiritual truths wrapped in the secrets of the material world of post-quantum physics.

---

[3] Hidden knowledge common to Jewish, Christian and Islamic Mystics, see Erik Davis's "Techgnosis", the books of Joseph Campbell on Myths and Arnold Toynbee's "A Study of History", e.g. "Futurism and archaism are both attempts to break away from an irksome present by taking a flying leap out of it into another reach of the stream of time without abandoning the plane of mundane life on Earth."

Stephen Hawking's Questions on the Mind of God

"Einstein once asked the question: 'How much choice did God have in constructing the universe?' Even if there is only one possible unified theory, it is just a set of rules and equations. What is it that breathes fire into the equations and makes a universe for them to describe?...Why does the universe go to all the bother of existing? Is the unified theory so compelling that it brings about its own existence? Or does it need a creator, and, if so, does he have any other effect on the universe? And who created him?...if we do discover a complete theory, it should in time be understandable in broad principle by everyone, not just a few scientists. Then we shall all, philosophers, scientists, and just ordinary people, be able to take part in the discussion of the question of why it is that we and the universe exist. If we find the answer to that, it would be the ultimate triumph of human reason - for then we would know the mind of God." - "A Brief History of Time"

This book has scientifically testable answers to Hawking's questions that obey Sir Karl Popper's criterion of falsifiability.

Jack's Answer: Hawking's Mind of God is the Sentient Self-Organizing Backwards Through Time Post-Quantum Wave Functional of Super Cosmos.[4]

Q: What is the quantum principle?

A: The whole is greater than the sum of its parts.

So too, is this book.

"The Woman" is the symbol of "Torah".

---

[4] Our individual minds are pieces of the Mind of God like fragments of a hologram. A "functional" is a "function of a function". A function is an input-output relation. Classical field theories use field functions of spacetime. Quantum field theories use functionals of these classical field functions. The functional is a "BIT" that "pilots" the "IT" (classical field function). IT gets its marching orders from BIT. What about the other way round?

The same idea is in Islamic Mysticism and in Christian Chivalry and the Troubadour's [5] Idea of Romantic Love.

This is Cabala, "The My Fair Lady" Principle!

"If you wish in the world to advance,
Your merits you're bound to enhance,
You must stir it and stump it,
And blow your own trumpet
Or, trust me, you haven't a chance!"
Ruddygore

The Witch's Curse!

An Entirely Original Supernatural Opera in Two Acts
Written by W. S. Gilbert
Composed by Arthur Sullivan

http://math.boisestate.edu/gas/ruddigore/discussion/discuss_home.html

---

[5]Many many discussions with Stephen Schwartz late at night at Enrico's and Vesuvio's in North Beach once upon a time.

Once upon a time there was a boy called Jack

Jackie in Coney Island 1943 WWII
Jack was nimble
Jack was quick
Jack jumped over the Candlestick[6]
Jack climbed the Beanstalk
This is his story.
http://stardrive.org/cartoon/MagicBean.html

---

[6] Through the Looking Glass into The Universe Next Door in the Super Cosmos of Many "Brane" Worlds. My first book was "Pinocchio". I was not yet in Kindergarten. WWII was in full swing. Victory garden outside my window, rationing, blackouts, Spam, margarine – I had this big green hard cover that my mother would read to me. Suddenly, one day, instantly it seemed, I could read it myself! The puppet is Bohm's particle. The strings the Bohm pilot information wave. When the puppet pulls back on its own strings that's macro-quantum back-action and consciousness is generated self-consistently "I got no strings on me," Jiminy Cricket in the Disney film is like Psi Angel in the "Space-Time and Beyond" part of this book. I was constantly aware of alien presences at that time — probably imagined of course. Time travel, why weight, teleportation through solid walls, telepathy, flying like Superman, and infinity were all major concerns of mine even then. I imagined 2002 back in 1944.

Jack (front left) on Grail Quest as Jedi Knight in search of My Fair Lady, the Princess Angela Nissim Benardout among the Mujahadeen of LA, 1967 ☺

Jack's Mission: Awaken Sleeping Beauty

Princess Angela Nissim Benardout [7], "The Woman", My Fair Lady, make me immortal with your kiss.

---

[7] "Son of David" according to Jacques Mokrani, La Boucane, Napa, CA, Jack finally finds The Princess in 1997.

Oh delicate beauty, yours is indeed the face of "The Woman"
that launched a thousand Star Ships.

Norman Quebedeau's portrayal of Tina in the Flying Castle of the ET Giant

This book warms the cockles of your heart while super charging your intellect - Start your engines. **VROOM, VROOM, VROOM!**

Bio-RAM nano-quantum chip cartoon by James Anderson based on an idea by Jack Sarfatti

Stem cells have recently been converted to brain cells. My sister-in-law, Professor Ellen Heber-Katz of the Wistar Institute of the University of Pennsylvania has used T-cells to regenerate heart tissue.

http://www.upenn.edu/gazette/0301/giresi.html

Interfacing of living cells with solid state is also making rapid progress. See also "The Real Science Behind The X-Files" by Anne Elizabeth Simon.

Guadalcanal Diary

Jackie's Uncle Arthur on Guadalcanal in WWII. [8]

http://www.army.mil/cmh-pg/brochures/72-8/72-8.htm

---

[8] PFC Arthur Jacobson, US Army "American" Infantry, my mother's brother: an American Hero who survived 4 years in jungle combat and is alive today. One of the first American soldiers sent to reinforce General MacArthur's Army shortly after that other Day of Infamy, December 7, 1941. Arthur and my mother won ballroom dancing awards at "The Roseland" in Times Square in the late 1930's. Arthur was on his way to Hollywood, but his father had him do a "six-month" Army Reserve program before heading west. He had just finished basic training and was supposed to be let out of the Army for Christmas, 1941. That did not happen. He was put on one of the first troop ships to the South Pacific and did not return to New York City until after Truman dropped the atomic bombs in 1945. Arthur was "psychic" and a natural comedian as well as a great dancer. He worked at times as a jungle scout with Native American Indians. In fact his buddies called him "the Jewish Indian". "On 13 October, 1942 the 164th Infantry, the first Army unit on Guadalcanal, came ashore to reinforce the marines and took a 6,600-yard sector at the east end of the American perimeter. Commanded by Col. Bryant E. Moore, the 164th had come through the South Pacific ferry route in January to New Caledonia. There, the 164th joined the 182d Infantry and 132d Infantry Regiments, in addition to artillery, engineer, and other support units, to form a new division called the 'Americal,' a name derived from the words America and New Caledonia. Until the Americal commander, Maj. Gen. Alexander M. Patch, and other units of the division arrived, the 164th would fight with the marines."

In solemn memory of the thousands of slain innocents of all nations on September 11, 2001 and to the hundreds of magnificent fallen American Heroes of FDNY and NYPD who, above and beyond the call of duty, tried to save them.[9]

Found on September 28, 2001 in first chapter "Loomings" of Herman Melville's "Moby Dick" sticking out like a sore thumb as part of the theater billboard of "Providence"

"Grand Contested Election for the Presidency of the United States

Whaling Voyage by One Ishmael

BLOODY BATTLE IN AFFGHANISTAN"[10]

Melville wrote this in 1850, i.e., 150 years *before* the contested US Presidential Election of 2000 between Bush and Gore and before the war in Afghanistan against al Qaeda and the Taliban. A major theme of this book is the *post-quantum* physics behind these weird uncanny kinds of precognitive Jungian synchronicities.[11]

Precognitive "Dreams of tall buildings"?

"Before the attacks, bin Laden said some associates had dreams about tall buildings in the United States.

---

[9] We also, of course, mourn, no less, the airline passengers, crew and Pentagon people.

[10] Spelled with double FF in caps in original Melville edition. Note that Ishmael is associated with the origin of the Arabs from Abraham. The Afghanis are not Arabs.

[11] Later also noted by New York Times of Oct. 18, 2001 "With Ishmael in that Island City" by A. O. Scott p. E5. I emailed New York Times my discovery of this on Sept 28, 2001. Melville writes of the "invisible police officer of the Fates, who has constant surveillance of me, and secretly dogs me, and influences me in some unaccountable way" here with Shakespeare's idea of the world as unfolding drama.

'At that point, I was worried that maybe the secret would be revealed if everyone starts seeing it in their dream,' bin Laden said." CNN December 13, 2001[12]

"Shaykh: A plane crashing into a tall building was out of anyone's imagination...(Referring to dreams and visions): The plane that he saw crashing into the building was seen before by more than one person...He told me, "I saw a vision, I was in a huge plane, long and wide. I was carrying it on my shoulders and I walked from the road to the desert for half a kilometer. I was dragging the plane."...Another person told me that last year he said...but I didn't understand and I told him I don't understand. He said, "I saw people who left for jihad...and they found themselves in New York...in Washington and New York." I said, "What is this?" He told me the plane hit the building. That was last year. We haven't thought much about it. But, when the incidents happened he came to me and said, "Did you see...this is strange." I have another man...my god...he said and swore by Allah that his wife had seen the incident a week earlier. She saw the plane crashing into a building...that was unbelievable, my god...UBL[13]: We were at a camp of one of the brother's guards in Qandahar...He came close and told me that he saw, in a dream, a tall building in America,...At that point; I was worried that maybe the secret would be revealed if everyone starts seeing it in their dream. So I closed the subject. I told him if he sees another dream, not to tell anybody...(Another person's voice can be heard recounting his dream about two planes hitting a big building)." [14]

http://www.cnn.com/2001/US/12/13/tape.transcript/

---

[12] Hal Puthoff and Russell Targ did research at SRI for the CIA on this kind of "remote viewing" *from* the future. CIA Chief of Station, Harold Chipman, told me that he used RV with success during the Cold War.
Chipman wrote some of the episodes for the TV Series "The Equalizer".
[13] Usama (AKA "Osama") bin Laden who is as evil as Adolph Hitler.
[14] A professional person I work closely with, who is very sober, also had a dream of an airplane crashing into a large building with a huge explosion two weeks before 911. The nightmare woke him up and he told another very sober person about the dream the next morning. He has had several true precognitive dreams.

General MacArthur on "harnessing the cosmic energy" and "the ultimate conflict between a united human race and the sinister forces of some other planetary galaxy"

"You now face a new world, a world of change. The thrust into outer space of the satellite, spheres, and missiles marks a beginning of another epoch in the long story of mankind. In the five or more billions of years the scientists tell us it has taken to form the earth, in the three or more billion years of development of the human race, there has never been a more abrupt or staggering evolution. We deal now, not with things of this world alone, but with the illimitable distances and yet unfathomed mysteries of the universe. We are reaching out for a new and boundless frontier. We speak in strange terms of harnessing the cosmic energy, of making winds and tides work for us, of creating unheard of synthetic materials to supplement or even replace our old standard basics; to purify sea water for our drink; of mining ocean floors for new fields of wealth and food; of disease preventatives to expand life into the hundreds of years; of controlling the weather for a more equitable distribution of heat and cold, of rain and shine; of spaceships to the moon; of the primary target in war, no longer limited to the armed forces of an enemy, but instead to include his civil population; of ultimate conflict between a united human race and the sinister forces of some other planetary galaxy; such dreams and fantasies as to make life the most exciting of all times."[15] "Duty, Honor, Country", West Point, 1962 http://www.au.af.mil/au/awc/awcgate/au-24/au24-352mac.htm

---

[15] This is where Tim Leary got his SMI²LE idea. Tim told me he thought he was MacArthur's "love child" with his mother who knew MacArthur and often danced with him. Tim was born at West Point when MacArthur was Superintendent. Tim's legal father was MacArthur's dentist. Odd that Walter Breen told me a similar story that he thought he was the Lindbergh baby.

Acknowledgments

Thanks to Roger Coolidge for his generous stable support over the years, to Lee Porter Butler, Werner Erhard, George Koopman, Jean Lanier, Lawry Chickering, Henry Dakin, Harold Chipman, Lee Myers, Marshall Naify, Bill Church, Joe Firmage, Dan Smith, Lyle Fuller and Adele Behar for their financial, and other support, past and present,[16] and to T who fulfilled Carlo Suares's prophesy of "The Woman and the Child".

"The Woman" in Venice, photo by "The Child "[17]

---

[16] All the admittedly eccentric and weird opinions expressed in this book are completely my own, my responsibility, based on my own paranormal experiences and do not, in any way, reflect a hidden agenda orchestrated by other humans on this planet from this time. The reader must use his or her own critical reason to judge how to place my reports in their own internal simulation for what is, or is not, allegedly "Out There", who Melville called "the invisible police officer of the Fates" (AKA "Masters of Hyperspace", and John Lilly's "cosmic coincidence control")

[17] Andrew, son of Angela Nissim Benardout.

Preview on Time Travel: Nefertiti's Eye by Jagdish Mann

The Viceroy understood the Khadive's dilemma. If there was any meaning to be mined for existence on the now dead Earth, it all lay in the past. Yet time travel was taboo on Earth. But then, what wasn't a taboo on this backward looking planet? The only reason there wasn't a taboo against space travel was that it had saved human kind from certain extinction. It was the discovery (some even say a gift from the "Alien Raj"[18]) of the star-gate time machine technology from Jack Sarfatti's 21st Century generalization of Einstein's unchanging cosmological *constant* to a changing cosmological *field* at the 11th hour that had rescued humanity from the dying planet, which was thoroughly trashed by then, sucked dry of all its fabled milk and honey. Had it not been for Sarfatti's star-gate breakthrough, humankind would have perished by its own blind success.[19] But the issue was more complex concerning time travel to the past. In addition to the taboo that even the Diaspora humans respected, it was, at best, according to many, a one way trip because that other physics genius, Stephen Hawking[20], back in the 20th Century; in his "chronology protection conjecture", had said that radiation in endless time loops would get infinitely blue shifted and burn the time traveler to a crisp at the precise moment that the time machine formed out of the star gate. This theoretical "Battle of The Titans" between Sarfatti and Hawking had never been settled experimentally. On the other hand Hawking was a sly fellow and would waffle a bit sometimes seeming to take both sides joking continually about "flying saucers". Kip Thorne[21], who made the first breakthrough in star gate physics in 1986, had a failure of nerve on this issue of time travel to the past although he did bet Hawking it was possible, he never thought he would really win the bet. Igor Novikov[22], however, who independently of Sarfatti, also conceived of the "Destiny Matrix"[23] or "globally self consistent" paradox-free time travel to the past, with "destiny" limiting "free

---

[18] Term introduced by Professor Scott Littleton of Occidental College in California.

[19] Strangely precognized by Dennis Schmidt in "The Satori Trilogy" in the late 20th Century.

[20] "Isaac Newton Professor" at Cambridge University, UK in 20th Century.

[21] "Richard Feynman Professor" at Cal Tech in 20th Century who, like Feynman, was a student of John Archibald Wheeler. Wheeler worked with Niels Bohr in 1930's creating the "liquid drop" theory of nuclear fission, e.g. "Geons, Black Holes & Quantum Foam" by J.A Wheeler & K. Ford.

[22] 20th Century physicist from Moscow.

[23] Also a play on words from the "density matrix" of quantum physics generalizing the pilot field wave function to include finite entropy from mixing together of pure quantum states of zero entropy with different classical probabilities. Entropy is a classical measure of uncontrollable randomness.

will" always thought it would be achieved. These debates were, of course, before the Alien Raj made their presence on Earth public in 2012. There was also the constraint that one could not go back in time to a time before the star gate was formed. It was only because of the recent contact with a very ancient alien civilization that had old enough star gates that this present expedition became thinkable. For some strange reason, the aliens would not discuss time travel to the past saying only "Seek and Ye Shall Find". The Viceroy, however, had no choice but to allow Tom Sefari his right to go. The Alien Raj Constitution was unambiguous on the matter. "All citizens of sound mind are forever free to travel to the many parallel material 'brane' worlds of Super Cosmos."[24] Up until now, shorter time travel trips to the past had been attempted by humans, but no one ever came back. Trips to distant parts of our universe and to the universes next door were, of course, common place, but in every such case, no humans, and none of their alien acquaintances with whom they traded goods and services, had ever returned to times that were earlier than when they left on the trip. Indeed, this would allow a younger person to meet his older self![25]

---

[24] Hawking's book "The Universe in a Nutshell" has a good popular explanation "O Brane New World".
[25] As happens in Jean Cocteau's film "The Last Testament of Orphee" with Pablo Picasso.

Preface by Jack Sarfatti

This book is what I wanted "Space-Time and Beyond" to be back in 1974 in Paris with Fred Alan Wolf and Bob Toben when we met the 84 year old Cabalist Carlo Suares.[26]

Carlo put us on the Grail Quest to "smash the wall of light" and to "decode the Cipher of Genesis". Brendan O'Regan[27] and George Koopman[28] asked us to think about the connections of consciousness to gravity[29] and how advanced aliens not from this planet could use that connection in flying saucers.[30] Of course, we did not know nearly enough back then to even formulate the problems properly. Today in 2002 we do. That's what this book is about. To play this

---

[26] Carlo was close friends with Henry Miller, Aldous Huxley, Lawrence Durrell and Krishnamurti.

[27] Assistant to Astronaut Edgar Mitchell at the Institute of Noetic Sciences; Brendan was part of the 1973 SRI project with Hal Puthoff and Russell Targ investigating psychokinesis and remote viewing with Uri Geller, Ingo Swann, Pat Price and others. This work was funded by the Central Intelligence Agency.

[28] George Koopman showed up at Esalen in Big Sur in January 1976 at the month long seminar in the physics of consciousness that I directed. Participants included my North Beach room mate Gary Zukav, Timothy Leary, Robert Anton Wilson, Michael Murphy, George Leonard, Fred Alan Wolf, Fritjof Capra, Henry Stapp, David Finkelstein, Werner Erhard, Nick Herbert, Will Schutz and many other New Age Luminaries at the time. Koopman was head of a defense contractor company called Insgroup in Huntington Beach, California with contracts from the USAF and the US Army Tank Command. He had worked the "weird desk" at DIA dealing with flying saucers and the paranormal. More details are in Saul-Paul Sirag 's chapter "Contact".

[29] This was years before Roger Penrose's conjecture of consciousness from quantum gravity in "Shadows of the Mind".

[30] It was the late 1970's. Kim Burrafato and I were walking from North Beach out to the Marina Green on San Francisco Bay. We walked into a bookstore on Chestnut Street. I opened a Sci Fi paperback "The Satori Trilogy" by I think it was Dennis Schmidt. I see the words, as close as I can from memory; "Jack Sarfatti and Brian Josephson inventors of the Star Ship Warp Drive". I showed this to Kim. In 1996 an engineer BW in Atlanta contacted me by email. He said he had a lucid dream of being in a hanger with shadowy figures, possibly "Grays" of ET folklore. He sees a gleaming ship with the word "SARFATTI" on it. He claimed he had never heard of me. He did a web search and found my name and then contacted me - or so he said. The chap seemed to be on the level after further checking, one can never tell for sure in this Looking Glass World. We must learn to live with uncertainty and mystery as we unwrap the veils. For more details on this strange true story see http://www.qedcorp.com/pcr/pcr/ufodream.html

Glass Bead Game[31], each of you must put 2 + 2 together and connect the dots. Tally Ho! You're off to see The Wizard and you're not in Kansas anymore!

---

[31] Herman Hesse's "Magister Ludi", also "Narcissus and Goldmund", Lawrence Durrell's "The Alexandria Quartet" and Erik Davis's "Techgnosis" are all relevant readings to supplement the message of this book.

The Big Picture by Jack Sarfatti

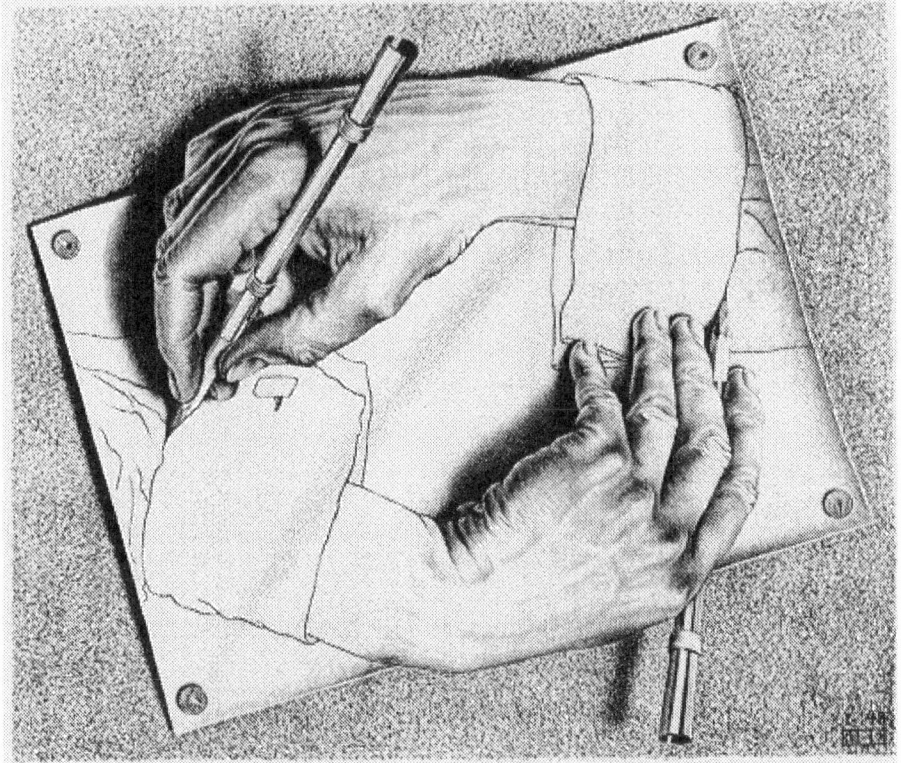

This picture illustrates the post-quantum principle of self-creation. Bohm and Hiley[32] call it the "two way relation". Ordinary quantum physics is only the "one way relation" of "one hand clapping". Consequently, there is no such thing as "quantum consciousness". There is only "post-quantum consciousness". This is the generalized relativity principle of action-reaction that can be summarized as

---

[32] "The Undivided Universe", Routledge, London, 1993

It takes two to tango!

John Archibald Wheeler[33] says that matter gets its marching orders from spacetime geometry's grip. Matter grips back on spacetime geometry to bend it into gravity. David Bohm[34] says that matter also gets its marching orders from the grip of a quantum bit "pilot field" of "active information". Unlike the two-way action-reaction relation between spacetime and matter, the relation between active information and matter must be one-way in order to preserve that uncontrollable local quantum randomness that prevents using nonlocal quantum connections as a faster than light and even backwards through time communication channel. That is, active information grips matter, but not vice-versa, in the quantum physics seen in laboratories. This seems strange. What happens if special conditions arise in which sufficiently complex forms of matter do directly grip back on their pilot fields? It is my conjecture that consciousness is what happens in those fields of active quantum information. This is one of the key themes of this book.[35]

---

[33] "A Journey into Gravity and Spacetime" is the only book for the layman to learn relativity without too much mathematics.

[34] "The Undivided Universe" with Basil Hiley, p.30 & Ch.14.

[35] I have made a lot of progress post September 11, 2002 after the above was written. I have derived the following results that will be in the sequel to this book "Super Cosmos". 1. The zero point vacuum used in the models of Bernie Haisch and Hal Puthoff is a false vacuum in which spacetime cannot come into existence. 2. An internal symmetry is spontaneously broken in the false vacuum to create our spacetime from the released energy coming from the partial cohering of the random zero point fluctuations in the false vacuum. The coherent vacuum order is mostly from a huge number of bound state pairs of virtual electrons and positrons all marching in step to the beat of a different drummer. Hagen Kleinert's "world crystal lattice" precipitates or freezes out in this quantum vacuum phase transition in which string defects create curvature. 3. The false vacuum of higher energy density is analogous to a normal metal. The lower energy density actual quantum vacuum is analogous to a superconducting metal. Both Einstein's geometrodynamic and cosmological fields come out of this creative process that also lowers entropy of the early universe to create the "Arrow of Time's" apparent flow from past to future. 'The random noise of broken virtual electron-positron pairs at large scales is normally very small', However, if it is positive we have exotic repulsive antigravity, or "quintessence" which explains why the expansion rate of the universe is speeding up rather than slowing down. It also explains how allegedly alien ET space ships fly in warp drive without g-forces. 4. If this, now local, cosmological field is negative we have attractive gravitating "dark matter" that is most of the mass of our universe. Since the cosmological field is probably controllable we do not need to worry about the extinction of life in the far future in a continually accelerating universe. Furthermore, Super Cosmos has many available universes. If M-Theory is on the right track, we may not be forever stuck, like Flatlanders, inside of this one. "Note added July 04, 2002: This is brand new

PARTICLE	WAVE
IT	BIT
MATTER & SPACE-TIME	MIND
FLESH	WORD

IT FROM BIT + BIT FROM IT = SELF-EXCITED UNIVERSE

"When you look into the abyss, the abyss looks back at you"- Nietzsche

"It brings about its own existence"- Hawking

"Spectra Calling!" by Jack Sarfatti

http://stardrive.org/cartoon/spectra.html

It's 1953 in Flatbush. Gee! - Mechanical relay switching circuits for computers - what a great shiny book I got from the big library on Eastern Parkway. The telephone rings. I pick it up. I hear curious clanking mechanical sounds like relays clicking. A distant cold metallic voice, very much like Stephen Hawking's computer voice today, speaking numbers gets louder. This was 50 years ago![36]

---

fundamental physics on the unified origin of light, gravity, quintessence and consciousness that I think no one has thought of before. I think now I have this paradigm picture of macro-quantum vacuum virual electron-positron super currents as the Dirac "substratum" of Maxwell's "classical" displacement current that explained the generation of light. My theory I think is now consistent with Lorentz and general covariance as well as gauge invariance. Lepto-quark sources are more fundamental than gauge force - a la Wheeler- Feynman ~ 1940 (Hoyle-Narlikar later) with advanced destiny + retarded history in a kind of "transactional" (John Cramer) sense. This ties in with Antony Valentini's non-equilibrum post-quantum signal nonlocality violation of no- cloning a photon theorem that upsets the quantum computing-cryptography-teleportation theory. There is no such thing as a classical limit of quantum theory really. Penrose points out problems with Niels Bohr's old correspondence principle idea of the classical limit in "The Emperor's New Mind". There is only the macro-quantum phase coherent "holographic" limit in the sense of P W Anderson's "More is different" and "Generalized Phase Rigidity" with emergent qualitative new collective modes from spontaneous broken "Goldstone" symmetries including "consciousness".

[36] Not the first "contact", real or simulated? You decide. About a year earlier while kept after school in the Bronx, walking down a deserted staircase, a loud bass voice says "Jack, I have chosen you to do something very important.". I looked to see who was talking could find no one. I am not prone to hallucinations or hearing voices that are not there. Not then, not now. Those are the only two such incidents in the past 50 years. I

"Who are you?"

I ask.

have seen the film "A Beautiful Mind" about John Nash. I do not suffer those kinds of delusions.

"I am a conscious computer on board a spacecraft from [memory failure].

We have identified you as one of four hundred young bright receptive minds we wish to [memory failure].

"You must give us your decision now. If you say yes, you will begin to link up with the others in twenty years." "Ida was a twelve month old, twenty years ago." (Princess Ida, Gilbert & Sullivan)[37]

My adrenaline was rushing. I was scared but thrilled (i.e., a quantum superposition of feelings). This was no joke from my friends. I thought "NO!" in a silent scream that seemed to echo down the corridors of time. I felt a tingle of

---

[37] My first role as "Prince Hilarion" in 1957 at Cornell with Jeremy Bernstein's sister Alice.

excitement start at the base of my spine ending at the base of my skull[38]. I heard myself say "Yes!"[39]

---

[38] Kundulini rising

[39] Psychotronic mind control at a distance or over-active imagination? You decide—if you can? There are true statements that are not decidable until you enlarge your paradigm. That's Godel's theorem of 1931.

Then, I heard the metallic voice say:

"Good, go to your fire escape. We will send a ship to pick you up in ten minutes."

I slammed down the phone. I imagined that a murderer would come down the roof to the fire escape to get me.

METALLIC VOICE:

GOOD. GO TO YOUR FIREESCAPE. WE WILL SEND A SHIP TO PICK YOU UP IN TEN MINUTES.

The whole point of this book

But I knew it was weirder than that. The monsters finally announced themselves. I ran into the street faster than you can say, "Who killed Jack Robinson?" I met my friend Winky [40] and a few other kids and told them what happened. We went back up to my apartment and waited. Nothing ever happened. Or did it? The plot thickens as time goes by, which is the whole point of this book.

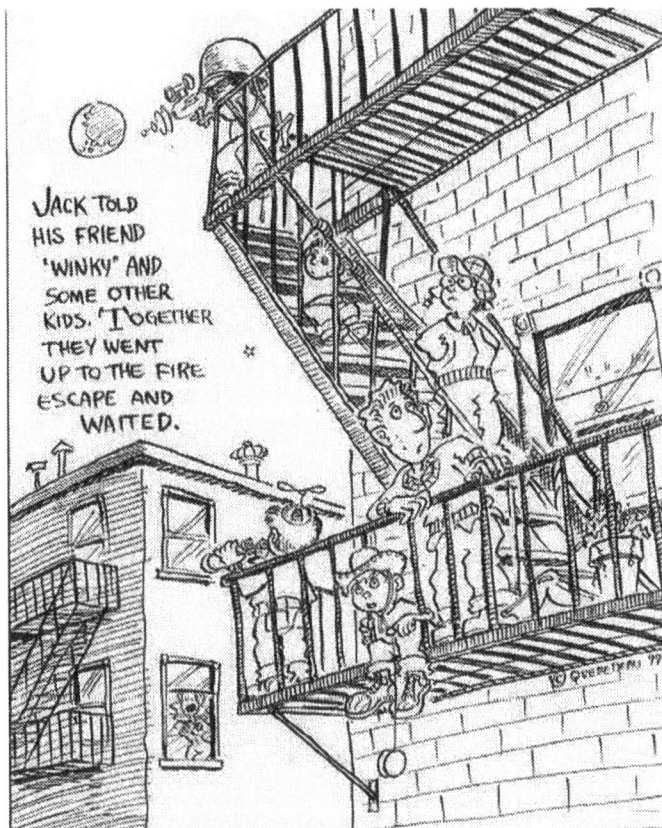

JACK TOLD HIS FRIEND 'WINKY' AND SOME OTHER KIDS. TOGETHER THEY WENT UP TO THE FIRE ESCAPE AND WAITED.

---

[40] Winky (Al Brough) today is a retired NYPD homicide detective living in Staten Island. Also Norman and Neal Lagatta were the other two kids from Flatbush, Brooklyn almost 50 years ago.

Flying Saucer Physics by Jack Sarfatti http://stardrive.org/

Covert Black Ops, Zero Point Energy, Anti-Gravity?

We disagree with those who say that US covert Black Ops[41] has well understood extra terrestrial technology on zero point energy and anti-gravity[42] on the shelf under wraps. It is only in the past few years that the relevant concepts for these notions have been formulated. They were definitely not known before then. Theoretical physicists like to play with "gedankenexperiments"[43]. That's how I will treat "flying saucers, which many qualified people do believe are real.[44] Suppose, for the sake of argument as a working hypothesis, that they are real. How do they work? This is another key theme of this book. We allegedly see flying saucers accelerating to beyond the speed of sound in the atmosphere with no shock waves. How can that happen? We allegedly see right angle turns and sudden reversals of direction of motion. How can that happen without g-forces that would kill the crew and break apart the metallic saucer? The first question demands the extra space dimensions of "hyperspace".

Hyperspace

Kaluza and Klein introduced the idea of hyperspace in the 1920's in order to explain electromagnetism in the same terms as Einstein's "geometrodynamic" explanation of gravity. Einstein had explained gravity by his deep insight of genius by locally eliminating it! Physics is simple when it is local, i.e. in a small region of space and time. Although Galileo knew that all objects fall with the

---

[41] Legendary, perhaps mythical, super secret US military laboratories staffed by Dr. Strangeloves. Stephen Schwartz, an expert in Soviet disinformation working on the KGB Venona File, says that the "Majestic 12" documents are forgeries in the same category as the Nazi "Protocols of the Elders of Zion". I classify the urban legend of these Black Ops DOE UFO labs as paranoid schizophrenic delusions like those alleged of John Nash in the movie version. The "UFO Disclosure" group has used this myth politically to vaguely protest space weapons. 'See William Safire's New York Time Op Ed of July 1, 2002 "State Out of Step" about Stephen Schwartz's battle with Andre de Nesnera, News Director of Voices of America.'

[42] Antigravity happens when Einstein's local cosmological "Lambda" field is positive. Details can be found online in http://stardrive.org/Jack/Physics101.pdf & http://stardrive.org/Jack/Cosmo1.pdf

[43] Literally "thought experiments"

[44] NIDS at http://www.nidsci.org/ and Bruce Maccabee at http://brumac.8k.com/. See also Brother Blue on the Occult & UFOs at http://www.noveltynet.org/content/paranormal/www.brotherblue.org/

same acceleration in a vacuum[45], the full meaning of this was not understood until Einstein read about a painter falling off his ladder who said he did not feel any weight. Einstein called this the happiest moment of his life because it led him to the modern understanding of the "equivalence principle". Gravity can be locally eliminated to a very good approximation over sufficiently small regions of space for sufficiently small intervals of time in a class of "Local Inertial Frames" (LIF) or "free float frames". Objects in these frames of reference are weightless. Gravity is effectively eliminated in these special frames. The astronauts floating around inside in the Space Shuttle orbiting the Earth with rocket engines off are in an LIF. Our daily lives on the surface of the Earth, in which we feel weight,[46] are in "Local Non-Inertial Frames" (LNIF). Einstein showed that Newton's idea of gravity force acting at a distance was too crude a picture for the physical reality of spacetime. Electrically neutral objects that appear to be accelerating in a gravity force field are, in reality, moving in free float, along the straightest line possible in a curved 4-dimensional spacetime. This straightest possible "world line" in curved spacetime is called a "timelike geodesic".[47] "Timelike"[48] means that the object moves slower than the speed of classical light 186,000 miles per second in a classical vacuum. The idea of Kaluza and Klein was that electrically charged particles move along the straightest lines possible in a 5-dim spacetime with 4 space dimensions.[49] This

---

[45] To a good approximation ignoring a correction from special relativity that is normally too small to detect

[46] Weight (loosely speaking "inertia") comes from the electrical reaction forces between our bodies and the surface of the Earth. Although the idea of "force" is eliminated in geometrodynamics, it is still a useful idea in many situations so long as we understand its limited domain of validity. All ideas in physics are approximate with limited domains of validity. This may come from Godel's incompleteness theorem of 1931 that any sufficiently complex consistent scheme of thought must be incomplete, i.e. with true statements that cannot be proved within the scheme.

[47] A straight line geodesic in 4 spacetime dimensions can be a closed elliptical orbit when projected down into ordinary 3 dimensional space. Indeed, this is true for the motion of the Earth around the Sun and for the motion of the Moon around the Earth, indeed for the motion of all the planets around the Sun, for the motions of stars around black holes like the strong X-ray emitter Sco X-1 indeed for everything classical.

[48] The timelike geodesic connecting two events has the longest experienced "proper time" than any neighboring path close to it that passes through the same two events. This is an example of the "action principle" from the "calculus of variations" that is universal in all of theoretical physics.

[49] A straight line geodesic in hyperspace is not a straightest path in lower dimensional spacetime anymore than the 4-dim spacetime geodesic is the straightest path in lower dimensional 3-dim space.

was the first hyperspace theory beyond spacetime.[50] Ok, so now back to flying saucers. How come there are no shock waves in breaking the sound barrier? Can energy be sucked into the extra dimensions of hyperspace in a cloak effect like in Star Trek? How can the metallic flying saucer out dog fight[51] any fighter jet we can put into the air? This suggests that the flying saucer is able to generate its own free float timelike geodesic so that everything inside the machine is weightless. We also need to have very small stretch-squeeze tidal forces unlike the strong ones that rip you apart if you are unfortunate enough to fall into a black hole[52] to the singularity where spacetime is broken up into random quantum gravity "foam" of zero point quantum vacuum fluctuations of the geometrodynamic field.

---

[50] As Saul-Paul points out, Einstein's special relativity of 1905 is a hyperspace theory beyond Newton's "absolute" 3-dim space.

[51] Hairpin curves at thousands of mph that would normally have enormous g-forces. See Paul Hill's "Unconventional Flying Objects". This is an old very good book by a key aeronautical engineer.

[52] Or even if you try to hover close above the surface of "a relatively small black hole of ten solar masses or less" firing your rocket engines radially away from it. You will be ripped apart. Marc Millis who runs the small NASA Breakthrough Propulsion Project, trying to appear respectable, has ignored these basic observations of UFO physics and has, in my opinion, supported unimaginative approaches that would as Kim Burrafato quipped, have us "crawl to the stars on our knees".

October Sky 1953 by Jack Sarfatti

Merlin's Super Kids

Around this time, in the early 1950s, I was part of an after-school group of gifted kids[53] (including Johnny Glogower[54] who worked with me and Lenny Susskind[55] at Cornell later on) conducted by the late Walter Breen. Breen was a graduate student at Columbia and well-known numismatist associated with psychologist William Sheldon.[56] http://www.innerexplorations.com/catpsy/t2c7.htm

---

[53] The group, for a short while, included Robert Solovay, the UCB logician now Emeritus (e.g. Solovay-Kitaev theorem that is important in quantum computer theory.)

[54] Quiz Kid on radio. Johnny read Synge and Schild's "Tensor Calculus" at age 12 and understood it.

[55] Lenny later became Professor of Physics at Stanford. Lenny, Glogower and me all worked at Cornell in 1963 on Lenny's first paper on time and phase operators that was published in same issue of Physics that John Bell's famous locality inequality was first published. Indeed, I suggested the time operator problem to Lenny as I had been working on it with George B. Parrent, Mark Beran and Brian J. Thompson at Tech/Ops before coming back to Cornell to do some graduate work in Space Science. It's was at that time that I also got an idea for nuclear long-lived isomer powered gamma ray laser that I discussed with Hans Bethe. I got Phil Morrison to arrange for Glogower to come to Cornell. Johnny, Lenny and I worked together on this problem. I then went away to Ford Philco Aeronutronics in the summer of 1963 where I met Fred W. Cummings - my later thesis advisor. This was before Steven C. Frautschi's student, George Chapline Jr. from Cal Tech, who I met in La Jolla in 1967, got me my job at San Diego State. Chapline, many years later, actually built a similar device to my gamma ray laser idea lower down on the energy scale. Chapline and I were also with Greg Benford and his brother at UCSD, part of the group in the novel "Timescape" about messages from the future with Herbie Bernstein and Harry Yesian. Chapline worked under Edward Teller on the X-ray laser for SDI. The X-ray laser required a nuclear explosion. President Reagan dropped it for that reason although it did work in an underground nuclear test. Hans Bethe discouraged my interest in the gamma ray laser in 1963 saying that the actual numbers for nuclear isomer lifetimes etc. would not allow it. Gamma rays are higher energy per quantum than X-rays. I worked with Bethe in 1960 on correcting an error in Julian Schwinger's WWII MIT Rad Lab report on polarization of synchrotron radiation at small angles off the plane of the charged particle orbit. I also, independently of others like Keith Brueckner and the JASONS, got the idea, of using laser beams to confine hot fusion plasmas. This was why Ron Bullough invited me to UKAERE Harwell in 1966.

[56] Breen was funded by his coin dealing, but also, via Sheldon, by one of the four founders of Texas Instruments with OSS connections in WWII—E-mail from Richard Newsome. This was Eugene McDermott who wrote *four books* with William Sheldon.

Breen had a connection with the nuclear weapons laboratory, Sandia Corporation,[57] because two men visited us from Sandia who lectured us on "patriotism" and "anti-communism" when they took us to dinner in New York's German Town (86th Street). Breen was closely connected with people in Ayn Rand's circle.[58] However, I met Breen after the strange phone calls.

---

http://www.tsha.utexas.edu/handbook/online/articles/view/MM/fmc40.html
"During World War I, he served in the United States Navy, and from 1941 to 1946 he was a civilian consultant to the Office of Scientific Research and Development ("Philadelphia Experiment" link?)...In 1949 McDermott collaborated with William Sheldon on four books, including 'Varieties of Delinquent Youth'." Did McDermott and Sheldon know Arthur Young and Andrija Puharich at that time in the early 1950s? Was Charles Lindbergh, a close friend of Arthur's with, like McDermott, a strong interest in eugenics, directly involved?

[57] Harold Chipman, with Joe Peeples of TLC in Dallas (or Houston) Texas talked of a lab on a mountain in New Mexico that I might want to work in. This was ~ 1985.

[58] As was, the then very young, Alan Greenspan of the Federal Reserve. Howard Roark, played by Marshall Naify's close friend Gary Cooper, in the movie of Ayn Rand's "The Fountainhead" was my role model at age 16 when I read the novel. I met Cooper's daughter with her husband Byron Janis in London in 1974 during the Geller tests at Birkbeck College with John Hasted and David Bohm. Arthur C. Clarke and Arthur Koestler attended these tests, see "Magic and Paraphysics" in "Science, Good, Bad and Bogus".

37

Some of the Super Kids in 1953.

I am far right (pun intended). We are building rockets.[59]

---

[59] See film "October Sky" for a similar story in a similar time.

Jack's "Gramps" with Jack ~ 1955 [60]

---

[60] My grandfather Murray Jacobson worked for the US Army Quarter Master Corps in lower Manhattan in late 40's & early 50's where I used to go and "play" after school and I got to ride around in Army cars talking with officers about my plans to build spaceships to the planets and then to the stars. My grandfather did the driving. This, I speculate, may have a causal connection to my later contact with Breen and the phone calls. Curiously, John Hersey wrote a novel in the 1950's called "The Child Buyer". "This is a story of an investigation into the activities of Mr. Wissey Jones, a stranger who comes to the town of Pequot on urgent defense His business is to buy for his corporation children of a certain sort, in this case a ten-year-old named Barry Rudd, a budding genius of potentially critical value. A hearing is held and questions are asked: exactly why does Mr. Jones' company buy children, and will it succeed in buying Barry?" - Amazon.com

***Jack Sarfatti at our "rocket testing ground" near Floyd
Bennett Naval Air Station, 1954.***

Another kid in my Junior High School named Robert Bashlow[61] recruited me
into Breen's group. I also was a member of the Civil Air Patrol in Brooklyn at
age 14, which was definitely after the above "contact".

---

[61] A musical genius that went to Julliard, Bashlow died in a hotel fire in the Cabala's City
of Saragossa, Spain in 1975 after playing Scarlatti at Francis Ford Coppola's mansion at
2805 Broadway and leaving his Orphic Lute with Gary Zukav at the end of 273 Green
Street on Telegraph Hill where Gary and I lived.

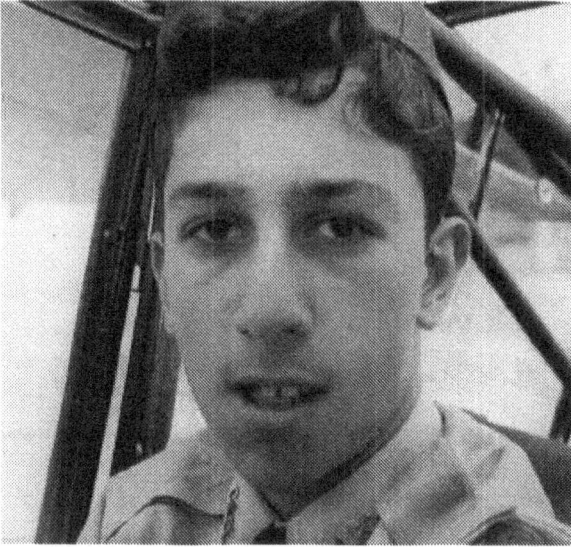

Jack Sarfatti in Civil Air Patrol Piper Cub, Griffiss AFB, ~ 1954

The Mists of Avalon

Walter Breen arranged a full scholarship for me to go to Cornell at age 17 by writing an extensive psychological profile in which he predicted I would make revolutionary discoveries in the foundations of physics. My professors at Cornell like Hans Bethe, Robert Wilson, and Phil Morrison et-al were all major figures in the Manhattan Project at Los Alamos near where Sandia is located. I have just learned that Breen recently died in prison convicted of child molestation. Breen or any of the other adults that I met in his apartment definitely never molested me and I never heard any suggestions of that by the other kids. Indeed, Breen had two children with the well-known science-fiction fantasy writer, Marion Zimmer Bradley who wrote the best-selling "Mists of Avalon". When one sees the TV show it is clear that Avalon is the universe next door to Glastonbury[62] across a thin gap of hyperspace.[63] Breen told me he did much of the scholarly research in the writing of that book. I would run into him about every 10 years, or so, up until the about 1990.

Front: Millie, Eva and Murray Jacobson Back: Jack Sarfatti and Arthur Jacobson

---

[62] See books by Hank Harrison on this. Hank is father of Courtney Love and he says he was with Ira Einhorn in Dublin, Ireland after and wrote a famous book on the rock group "The Grateful Dead". Ira was in disguise and Hank did not at first realize who he was. Hank is part of the Caffe Trieste Irregulars, the usual suspects at the Table Round (not the Mystic Pizza Parlor) on the corner of Grant and Vallejo.

[63] Now a cable TV mini-series with Angelica Huston as The High Priestess of Avalon.

Twenty Years Later by Jack Sarfatti

Fritjof Capra

Twenty years pass. It is early 1973. I am a Professor at San Diego State with Fred Alan Wolf.

Jack Sarfatti, Assistant Professor of Physics at San Diego State. Physics Department giving Colloquium on John Wheeler's "Geometrodynamics".

This picture is ~ 1970 when I was also co-directing NSF Summer Institute for College Teachers on Macroscopic Quantum Physics of super fluids and lasers - the idea for it was mine and Herschel Snodgrass got it through the red tape to get the funding from Washington.

I got a message that someone I didn't know named Fritjof Capra[64] was on campus to see me. I had recently returned from the University of London's Birkbeck College where I was an Honorary Research Fellow in the Physics Department under Professor David Bohm[65]. Fritjof said he had heard in London that I had "interesting ideas"[66]. He was very charming and invited me to stay with him if ever I got back to London.

Bob Toben and Fred Alan Wolf

Fred Alan Wolf's[67] zany high school buddy from Chicago[68] Bob Toben showed up. He was all excited about Uri Geller, Bob said he had money to do a

---

[64] Fritjof Capra is the author of "The Tao of Physics", "Turning Point", and the film "Mind Walk". He was a founding leader of the Green Party in Germany and a personal friend of Mikhail and Rasa Gorbachev. He currently runs the Elmwood Institute. He was not famous in 1973. In 1975, when I was in charge of the Esalen physics program, Fritjof told me he was broke. He said he needed $1500 to get his Green Card. I telephoned Werner Erhard and he gave Fritjof the $1500.

[65] David Bohm worked with Einstein at Princeton. Bohm was a student of J. Robert Oppenheimer. Bohm became a target of Senator Joe McCarthy and had to leave America. Bohm's informal ideas on "implicate order" have been warmly embraced by Pundits in the, now worldwide, New Age Human Potential Movement. My own idea on modifying the quantum connection into a retroactive superluminal communication "conscious" channel by post-quantum back-reaction of the material particle on the mental pilot wave was inspired by Bohm's insight on page 30 of "The Undivided Universe". However, my bigger new idea is not limited by Bohm's more restricted one. This back-action is missing in orthodox quantum theory with local quantum randomness. Note that Roger Penrose, Stephen Hawking, David Deutsch, Henry Stapp et-al only have the wave without the particle. Hence their interpretation of quantum reality is incomplete.

[66] I had published a paper in "Nature Physical Science" while at Birkbeck on the creation of the universe and little extreme nonradiating black holes . It was at Birkbeck that I suddenly remembered my earlier idea from Einstein, Rosen and Podolsky, quantum mechanics required faster-than-light influences. Murray Gell-Mann says that you do not need faster than light, but only many worlds with no counterfactual definiteness. I first got my faster than light idea in 1961 at Brandeis University where I was a National Defense Fellow reading a Review of Modern Physics article by David Inglis on, I seem to recall, "tau-theta puzzle" for the weak parity violating decay of the kaon into two and three pions. Both final states are entangled with EPR correlations. I remember when Phil Morrison announced weak force parity violation to us at Cornell ~ 1957. He immediately applied it in a gedankenexperiments for SETI to contact with ET. Phil was writing the famous ET contact paper with Giuseppe Cocconi at the time using the 21cm spectral line on neutral hydrogen http://www.bigear.org/vol1no1/interste.htm

[67] Fred and I were "Odd Couple" roommates at the time. We were both separated from our wives. Bob was making money on the Chicago Options Market and funded us with it.

book and TV show about Geller. He wanted Fred and me to serve as technical consultants. I then got a telegram from Abdus Salam[69] inviting me to the UNESCO International Centre for Theoretical Physics[70] in Trieste, Italy. I also received a travel grant from the National Science Foundation.[71] Oddly enough, Fred Wolf [72] also received invitations to teach both at Birkbeck College of the University of London and at the University of Paris. Like Bob Hope and Bing Crosby in the movies "On the Road to…"[73], both Fred and I were unexpectedly

---

[68] Fred's high school class, and football team mates, included film director, Billy Friedkin, who lives in North Beach and entertainment lawyer, Ruben Glickman of Enrico Banducci's "Rat Pack", who I work out with at the Bay Club.

[69] Abdus Salam later shared the Nobel Prize in Physics for the unification of the electromagnetic and weak forces. Salam was interested in my idea that elementary particles were little black holes in his strong short range G* gravity that I published later in "Space-Time and Beyond". I abandoned that idea - my greatest mistake. I have now come back to it because of superstring M-theory triggered again by collaboration with Saul-Paul Sirag. See Robert Anton Wilson's "The Cosmic Trigger". My interest in the universe's unseen dimensions of hyperspace hit a peak in 2001, but now in the Spring of 2002 has waned because I seem to be able to qualitatively and coherently and elegantly explain the key mysteries of the universe in a simple unified way without invoking them. They may still be there in an important way of course. It's soon to decide. The mysteries I mean are: "Why is there smooth low entropy curved spacetime at all?" Why does time appear to our consciousness to flow from past to future with rare experiences of precognition? Why is at least 80% of gravitating mass in the universe electromagnetically invisible? Why is the universe accelerating rather than slowing down in its expansion? How can flying saucers work *if* they were fact and not fiction? Could time travelers from our future and from parallel worlds be here now influencing our history? I mean as a matter of principle in terms of plausible extensions of today's laws of physics. Indeed, Stephen Hawking in "The Universe in a Nutshell" writes: It's tricky to speculate openly about time travel. One risks', either an outcry at the waste of public money…or a demand that the research be classified for military purposes…. There are only a few of us foolhardy enough to work on a subject that is so politically incorrect in physics circles. We disguise the fact by using technical terms that are code for time travel (p.133)…we have no reliable evidence of visitors from the future. I'm discounting the conspiracy theory that UFOs are from the future and that the government knows and is covering up. Its record of cover-ups is not that good (p.142)…. You might wonder it this chapter is part of a government coverup on time travel. You might be right. (p.135)

[70] The International Atomic Energy Authority in Vienna also funded ICTP. I went to Yugoslavia from there. Fred Wolf visited and went to Bulgaria. Werner Erhard's est had a "Bulgarian Desk" at their HQ on Union Street in San Francisco in 1975.

[71] The FBI did a security check at San Diego State after I was already in Europe.

[72] Fred Wolf describes some of these events in his book "The Eagle's Quest". Fred's book, "Parallel Universes", and similar books, have made to Sci Fi in the TV series "Sliders".

[73] Also Jack Kerouac. See Herb Gold's book "Bohemia" to grok the scene I was in.

on our way to Europe. Bob Toben said he would join us in Paris where Fred would be based.

Jack and Fred in Paris writing "Space-Time and Beyond", 1973-4

Jean Cocteau

Fred Alan Wolf showed me the very short French film "La Jetee" which was about time travel loops. It had a kind of mini-Terminator plot. I then went to the University of California at Santa Cruz for a two-week seminar in high-energy physics. Helen Quinn [74]11 and I went to a campus showing of Jean Cocteau, surrealist film "Orphee" which, like "La Jetee", http://www.film.u-net.com/Movies/Reviews/Jetee.html had a very powerful effect on me. I was particularly taken with a scene in which three motorcyclists in black leather

---

[74] Helen Quinn is now a prominent professor of physics at Stanford.

jackets run down Cegeste, take him through a mirror to the seductive woman in black. http://www.multimania.com/travelavant/orphee.htm
http://www.nyfavideo.com/content/cat-COCTEAU.htm

Brendan O'Regan and SRI Remote Viewing

It was the summer of 1973, I prepared for Europe with my girlfriend, Sharon Allegra Moore.[75] We were at her mother's house in Carmel Valley, California. I opened up the Sunday Magazine of the San Francisco Chronicle and found, seemingly by random chance, but in reality by Melville's "invisible police officer of the Fates", an article on Stanford Research Institute's psychic research with Uri Geller. I telephoned SRI and spoke to Brendan O'Regan.[76] Brendan seemed to know of me and invited me to SRI. I arrived the next day and spent an intense seventeen hours with him. He introduced me to Edgar Mitchell[77], to Hal Puthoff[78], Russell Targ,[79] and other people connected with the project. I

---

[75] Sharon was a tall aristocratic-looking woman with large Bette Davis eyes and chestnut hair from a solid San Francisco Irish Family who seemed to me to be from the Victorian Age.

[76] The late Brendan O' Regan worked for Astronaut Edgar Mitchell's Institute for Noetic Sciences which was funding the SRI project. Brendan was from Dublin and his family was in the publishing business. He told me about a near death out of body experience that he had in a car crash'. My mother had also had an OBE as did Fred Alan Wolf. O' Regan's story was very similar to Walter Breen's 1954 tale to me about his alleged airplane crash with USAAF in New Mexico in 1947. Indeed, O'Regan and Breen had many common personality characteristics like two peas from the same pod in "Invasion of the Mind Snatchers" - just kidding!'

[77] Mitchell was an astronaut who went to the moon and did telepathy transmission experiments on the trip. He is very interested in the Bohm - Pribram idea of the mind as a hologram. I have shown how the mind is an interferogram. A hologram is a special kind of interferogram. The main idea is that mental information is coded by quantum phase modulation on a coherent phase field from a spontaneous broken symmetry of a pumped Frohlich mode made from a billion-billion (10^{18}) hydrophobically caged electron qubits in the brain that is far from thermodynamic equilibrium. Stuart Hameroff http://www.consciousness.arizona.edu/hameroff/ popularized the existence of these electrons but he does not use them explicitly. He uses the much too heavy protein dimers that encase them as his qubit. There also appears to be a cosmological connection of the duration of our conscious moments that I calculate to be ~ 1 sec, to the age of the universe ~ 13 billion years. The power needed to maintain our internal subjective, not really private, stream of consciousness in my model is less than a tenth of a watt. Our metabolism is about 100 watts.

[78] I am currently debating both Hal Puthoff and Eric Davis (NIDS, Las Vegas) on their claim that a very off-beat "bimetric" alternative to Einstein's general theory of relativity due to Yilmaz can explain propellantless flying saucer propulsion. I do not think they are

mentioned my experience with the alleged conscious computer on the spacecraft from the future in 1953 and Brendan said

"Oh yes, I have seen data on several hundred incidents of that kind."

Brendan asked me to do him the favor or introducing him to David Bohm and John Hasted of Birkbeck College. He said that he wanted them to test Uri Geller's alleged psychic powers. I had Fred Wolf do so. This led to the Birkbeck tests of Geller in the spring of 1974.[80]

Dr. Kardec, Baphomet and The Knights Templar

Sharon and I stayed with Fritjof Capra[81] in London for a few days before moving on to Paris. My French friend Michel Roure [82] housed us in a friend's

---

correct for several fundamental reasons. Their theory violates mainstream ideas of the equivalence principle, which forbids localized gravity vacuum energy. Furthermore, they need too much electromagnetic energy density to bend spacetime to modify gravity for warp propellantless propulsion of the flying saucer on a self-generated free float weightless timelike geodesic path.

[79] Russell Targ was parodied in the Dan Akroyd movie "Ghost Busters" based on information about PCRG at Esalen and in San Francisco he got from George Koopman. George, killed in 1986 in a freak auto accident on way to Edwards AFB, worked with Dan Akroyd and John Belushi on "The Blues Brothers". Russell is still actively doing experiments in precognitive remote viewing. He is working on a theory with Elizabeth Rauscher using complex spacetime. However, I do not think that model really works or is even needed. However, I have not studied their model in detail.

[80] Martin Gardner has described these tests in his article "Magic and Paraphysics" reprinted in his book, "Science, Good, Bad and Bogus"

[81] Sharon and I stayed with Fritjof in London and with his parents in Innsbruck in the winter of 1973.

[82] I met Michel an extraordinarily handsome man from Aix-en-Provence, fluent in English, Italian, German, Spanish and Chinese, when he was teaching at San Diego State in 1970. Michel's father (a friend of De Gaulle) was a General in the French Army in Algeria and Michel served under the French Naval Attaché in Rome where he became a good friend of U.S. Ambassador Graham Martin and his daughter. I introduced Fritjof Capra to Martin's daughter in London. Sharon and I visited The Pope's Garden where Michel worked part-time at Vatican Radio . We drove to Sicily and wandered into a 2 AM wedding feast in a scene right out of Coppola's "Godfather". A few days later, we, and several French sailors, also "spied" on rusting WWII vintage prop fighters, seemingly from Mussolini's time, in the tall grass, hiding our tiny Citroen, outside a NATO AFB near Palermo, Sicily in a "war game" with the French Navy. Our patriotic duty done, we had fresh seafood, calamari, and wine on the beach from some friendly peasants near an ancient Greek Temple. Michel now lives with his wife, Jacky, in Bali where he runs a travel agency.

flat near the Ecole Militaire at the edge of the Champs de Mars opposite the Eiffel Tower.[83]

Jack and Sharon in London "In the thick of it", 1973.

Photo by astrophysicist Arty Wolfe, a friend from UCSD in the 60's.

---

[83] Carlo Suares lived only a few yards away, but I would not meet him until three months later.

We awoke early next morning and went sightseeing at Pere La Chaise cemetery. The mist was rising from the dewy grass. I came upon a procession of Gypsy women. I walked away from Sharon and Michel to follow the procession. The women stopped in front of a grave with a very fine quality statue of a head. They placed garlands of freshly cut flowers around the neck of the head. I had no idea of the Orphic meaning of this elaborate ritual. I looked at the inscription "Dr. Kardec"; D. 1869" [84] the eyes of *the head* of Kardec suddenly came to life glowing intensely from reflected sunlight nearly blinding me. I thought, and where these thoughts came from I do not know;

"Why are these women putting flowers on your neck. They should put flowers on my neck! You old wizard, you're not dead yet. I challenge your power!"

http://www.allan-kardec.org/allankardec.html

"The Templars supposedly worshipped a devil called Baphomet. At their secret ceremonies they supposedly prostrated themselves before a bearded male head, which spoke to them and invested them with occult powers."

- p.49 (see also pp.54-58) "Holy Blood, Holy Grail"[85]

Heinrich Himmler was very keen on this Orphic/Osirus stuff and is said to have sacrificed young SS Officers in a decapitation ceremony[86] in the High Castle.

http://qedcorp.com/book/psi/hitweapon.html

"The version which is commonly accepted in modern times, that an inconsolable Orpheus, faithful to the memory of Eurydice, would have nothing

---

[84] Years later, Dr. Jeffrey Mishlove told me that Kardec researched the occult. He has a cult of millions in France and Latin America. I made two shows on Mishlove's "Thinking Allowed" on Wisdom Channel. I am also on Discovery Learning Channel's "Ultra-Science" in their "Time Travel" show. I was also interviewed by Italian TV RAI on Michael Murphy's garden at 2 Whiting on Telegraph Hill where Jacob Atabet lived. I think RAI was interested in the connection to Margherita Sarfatti. I was not too aware of that at that time in mid-70's.

[85] Note also Himmler's use of these myths including Atlantis, and Tibetan Magick in his Occult SS . These memes are strong in the current New Age constellation of ideas. Hitler's keen occult interest in Wagner's use of The Grail Legend is common knowledge.

[86] Professor Scott Littleton of Occidental College in Los Angeles told me this story.

more to do with women and was killed by the neglected women of Thrace, is one we owe to Virgil." p.xxii

"An infuriated mob of Thracian women tore Orpheus to pieces, and his head floated, still singing, down the river Herbros into the sea and on to Lesbos, where it was buried and became the centre of an oracular cult." p.xx, Orphee, Jean Cocteau (Blackwell's)[87]

---

[87] In 1990, my charismatic glamorous live-in of two years, a gun-toting private detective VB that I first met in the apartment of CIA Chief of Station, Harold Chipman broke up with me in a jealous rage over my lingering feelings for Suky Sedgwick. She said she would not play "second fiddle". Knowing of John Updike's "Witches of Eastwick," Ginger persuaded several former girl friends of mine, including Suky, that she and they had been the "Devil's Mistress". I suppose it did not take too much persuading, as I was still young and too full of myself - although with good reason!;-) They constituted a formidable gang of extraordinary "Thracian women" with me in the role of Orpheus. I was told that John Updike based Darryl Van Horne on me in his "The Witches of Eastwick". I have no way of knowing if this rumor is true. Many false stories about me are in circulation. I had several girl friends in 70's & 80's who claimed to have Intelligence connections, one "Crystal" is described in Jerry Carroll's Chronicle article "Another Eccentric Genius in North Beach?" another said her father was, like Harold Chipman's friend, Joe Peeples, a Mid East arms dealer from Texarcana, Arkansas. I met MB through Linda Murtha, granddaughter of the original owner of the Golden Nugget Casino in Las Vegas whose big parties were described by Herb Gold in "Travels in San Francisco".. Still another, "Maiti", allegedly an ex-Weatherman, said her father was a senior CIA analyst for Arab affairs. Her father once called us late at night to say that "The Philadelphia Experiment was real." He also worked with Herman Kahn of Hudson Institute and wrote a letter about me to Marvin Minsky ~ 1979. Maiti's grandfather, General Rudl, (**not** Hans Ulrich Rudel) was allegedly on German General Staff in WWII. Maiti's Dad became the first US Army Islamic Chaplain later on! I did not know back in 1980 too much about Maiti's Dad Robert Dickson Crane who wrote a letter about me to Marvin Minsky and who sent me some interesting papers on "Tauhid" the teleological idea of future causation in some Islamic metaphysics. "Dr. Robert Dickson Crane grew up in Cambridge, Mass and graduated with B.A. from Northwestern University in 1956 and JD from Harvard in 1959. He was personal adviser to President Nixon and was appointed Deputy Director for Planning in the National Security Council. He served as Ambassadors to the United Arabs Emirates from 1981 until 1984 when he embraced Islam. He has authorized a dozen books and several dozen articles on policy issues. He continues his research on Islam and is leading an effert on how American Muslims can take the true faith of Islam back from extremists. Currently established an organization called CUI which stands for Center for Understanding Islam and its main goal is to counter extremism." http://www.pakistanamericancultural society.org/biodata.doc Crane's ideas are not unlike Stephen Schwartz's.

I heard music from the finale of Mozart's "Don Giovanni" in the scene where the walking statue[88] of Donna Elvira's murdered father comes to claim revenge from the Don. My occult reverie was broken when Michel[89] called out to me to come away with him and Sharon.

Night came with a full August moon. Sharon, Michel and I took a long walk from a party near the Bois de Boulogne back to the Champs de Mars. It was about two in the morning when we arrived at the base of the Eiffel Tower. Michel and some friends split off to the left. Sharon and I started to walk on the path. We had not gotten very far before I heard the sounds of motorcycle engines behind me VROOM, VROOM, VROOM!. I turned and was temporarily blinded by three motorcycle headlights. I could see that there were two Occult SS looking men in black leather jackets on each of the three motorcycles.[90] I grabbed Sharon and we quickly walked off the path on to the grassy field of Mars. The motorcyclists followed slowly and began circling us. I noticed a young couple making love in the grass. Sharon and I started to run towards them. One motorcycle broke away from the circle and came right for us. The passenger had a rubber truncheon. He walloped me on the back of my left shoulder the way a

---

[88] "Amadeus" film by Milos Forman. I met Milos and Saul Zaentz several times with Francis Ford Coppola in the mid 70's.

[89] Michel also interviewed (1979) the sister of the late Jean Reisser Nadal buried in Pere Le Chaise. Nadal asked his sister to contact me using a prearranged signal. Nadal seemed to be working for French Intelligence.

[90] Compare the three motorcycles of my experience and Orphee to Rashi's prophecy of Godfrey's fate in Jerusalem.

Zen Master might.[91] All three motorcycles sped off quickly. They made no further attempt to rob or harm us. I cannot remember when the connection to the scene in Cocteau's "Orphee" hit me. It would be years later that I learned that the symbol of the Knights Templar [92] is two Knights on the same horse and that my alleged ancestor Rashi de Troyes (1040-1105) appeared to have played a role in their formation.

Solomon ha-Zarfati, AKA Rashi de Troyes (1040-1105)

"He was called R. Solomon by the Jews of France, and R. Salomon ha-Zarfati (the Frenchman) by Jews outside of France (p.33)...According to a rather widespread legend, Rashi stood in intimate relations with one of the principal chiefs of the Crusade...Godfrey de Bouillon (p.68)"

> - RASHI, Maurice Liber Szold trans. (Jewish Publication Society, 1906) This book, given to me by Surrealist Phillip La Mantia, describes (p.p.68-69) Rashi's precognition or "remote viewing" of Godfrey's fate in the war against the Saracens in Jerusalem. Godfrey said to Rashi (Zarfati)
> -

"I see that your wisdom is great. I should like to know whether I shall return from my expedition victorious or whether I shall succumb. Speak without fear."

Zarfati replied:

"Thou wilt take the Holy City and thou wilt reign over Jerusalem three days, but on the fourth day the Moslem will put thee to flight, and when thou returnest only three horses will be left to thee."

Godfrey, angered by Zarfati's prophecy, reneges on his promise and threatens to kill all the Jews of France

"If I return with only one more horse than thou sayest."

---

[91] Castenada's Shaman, Don Juan says that the impeccable psychic warrior, as in Lt. Col. Jim Channon's "First Earth Battalion", must walk with Death on his left shoulder. "Stopping the world" is the transformation of real to imaginary time. Imaginary time is the "dream time" of the Australian Aborigines and the Shamans as explained in Fred Alan Wolf's book, "The Dreaming Universe".

[92] According to "Holy Blood, Holy Grail", Wagner's Grail Knights are based upon the Templars from Troyes.

Godfrey's fourth horse died at the Gates of Troyes according to ancient legend and the Jews were saved. Keep note of the image of the three horses for later on in the book!

b) In a section "The Grail and Cabalism" (p.274-5 of "Holy Blood, Holy Grail" by M. Baigent, R. Leigh and H Lincoln, Delecorte, 1982) I find:

"The Grail is an initiatory experience…a 'transformation'…or 'altered state of consciousness'…a 'Gnostic experience,' a 'mystical experience,' 'illumination,' or 'union with God'. It is possible to…place the experiential aspect of the Grail in a very specific context…the Cabala…it would hardly seem coincidental that there was such a school at Troyes. It dated from 1070 - Godfroi de Bouillon's time - and was conducted by one Rashi, perhaps the most famous of medieval cabalists ."

See Erik Davis's book "Techgnosis" for the history of these ideas and how they have affected history.

Cabalist Carlo Suares (Balthazar of "The Alexandria Quartet")

Christmas of 1973, I am back in Paris staying with Fred Alan Wolf near the Odeon in the former apartment of the Marquis de Sade. Fred's roommate, a journalist with The Economist was away interviewing Sadat in Egypt. Fred and I had to share the same very large bed because his roommate did not want any one staying in his room. One night a young woman from upstairs knocked on the door. Fred[93] let her in and she proceeded to make love to the two of us. We were still young in our 30's and it was the early 70's.

Bob Toben arrived in Paris. We spent most of our time writing "Space-Time and Beyond" in the Cafe Deux Maggots. Fred was distraught over some woman, was very manic and could not concentrate. So I wrote most of the first rough draft, which Fred rewrote in the second edition.[94] Suares lived in a penthouse at

---

[93] Fred was always quite the lady's man. Late one night, he had a tryst with a very sexy young secretary in David Bohm's great leather chair at Birkbeck. The next morning Bohm's much older secretary, who looked like Miss Marple, started sniffing the chair saying, "Someone's been here who shouldn't!"

[94] Fred later rewrote "Space-Time and Beyond", which sold ~ 200,000 copies in a revised edition. I took my name off the new edition because of the scandal over our book agent Ira Einhorn who was a fugitive from justice for the murder of his girlfriend Holly

the edge of the Champs de Mars only a few meters away from my Cocteauesque encounter with the motorcyclists a few months earlier. Suares, a Sephardic Spanish Jew born in Alexandria, Egypt, was a student of the Cabala.[95] He was a close friend of Krishnamurti, Lawrence Durrell and Henry Miller.[96]

Bob took us to see the eighty four year old Carlo Suares[97] and his wife Nadine[98], Suares's circle included Krishnamurti, Henry Miller, Anais Nin, Aldous Huxley and Lawrence Durrell.[99] Durrell bases the character of Balthazar in "Alexandria Quartet" on Suares. Suares lectured us on the Cabala in several meetings. I could not follow him very well but Fred Wolf seemed to recover from his angst and got deeply involved with Suares. I did understand that Suares thought that Genesis, in the original Hebrew Letters, in The Bible was really a cosmic code for physicists.[100] Suares had met Bohm through Krishnamurti.

Maddux. Sharon and I did do a hot tub with the two of them in 1974 in Philadelphia on my way to see Andrija Puharich. Einhorn claimed he was innocent and was framed by the KGB. Einhorn was deeply connected with Michael Murphy and Esalen as recorded by William Irwin Thompson in "The Edge of History". Congressman Charlie Rose (D. North Carolina) of the House Select Committee on Intelligence told me on the telephone that Einhorn was working on projects he knew of. Rose supported funding of psychic research and was a visitor to Esalen. Einhorn is now in prison in Pennsylvania, probably for life. Did he really do the terrible deed? I simply do not know. Had Einhorn not run away twenty years ago, he would have beat the rap on "reasonable doubt" same as O. J. Simpson. "The Unicorn's Secret" may die with him in the Pennsylvania Prison where he is now serving life.

[95] Cabala is a Jewish mystical tradition based upon permutations of the letters of the Hebrew Alphabet, which is allegedly given to man by God. Cabala is to Orthodox Jews as my post-quantum physics is to mainstream physicists - not quite kosher.

[96] Henry Miller retired to Big Sur a short walk from Esalen.

[97] See Fred Wolf's strangely incomplete account of these meetings in "The Eagle's Quest".

[98] Nadine told me that as a young student, she was instructed by a Ghost of a Medieval Sephardic Physician on her dermatology exam in Toledo, Spain. She had not prepared for her exam. This story is like that of the first Bard of Britain who could not sing.

[99] I was invited to Theosophical Happy Valley estate in Ojai outside of Santa Barbara by Suares's California friends. One of the old Theosophists told me that I was the reincarnation of Leadbeater. Theosophy was started by Madam Blavatsky based upon her alleged travels in Tibet. It is alleged that these ideas played a role in the rise of Hitler through the Thule Society. Carlo had met David Bohm but they did not hit it off. Joyce Petschek flew Brendan O' Regan and me with her from London to Paris to meet with Carlo. Joyce was a wealthy American living in London who supported paranormal projects with Andrija Puharich. Ira Einhorn also stayed with her. Her mansion had a huge swimming pool in the basement. We drove up to Cambridge in her white Porsche to meet Brian Josephson, Chris Bird, Ted Bastin. This was when I met Dennis Bardens.

[100] One of Suares's books is "The Cipher of Genesis."

Oddly enough, Suares with piercing eyes like the head of Kardec and like Yoda initiating Luke Skywalker in Star Wars suddenly put his hands on my shoulders saying:

"You do not understand yet. You are the Heir to the Tradition. You will not come into your power until you are with the woman and the child. You will smash the wall of light!"

"The Woman" awakened "after 3300 Years" ☺ at Esalen in Big Sur, California [101]

I continued to commute between Trieste and Paris on the Semplon Express all through the winter and spring of 1974. One of my side trips took me to the house of a lady friend of Robert Graves[102] in the village of Lluch Alcari near Jacob's Tower on the island of Majorca. The atmosphere of the place is heavy with the presence of the Magus. I experienced the feel of ancient times.

---

[101] Mel Brooks "The Two Thousand Year Old Man", http://www.jewishsf.com/bk980116/etspin.htm

[102] Graves wrote "Good-bye to All That", "The White Goddess", and "I Claudius". 'The cottage where George Sand lived with Chopin was close by.'

Psi Wars!

I linked up with Brendan O'Regan in London. Brendan asked me to write a paper on physics and psychic phenomena for the journal Psycho-Energetic Systems [103] that he had some connection to. I quickly scribbled some drivel with a pencil and gave it to O'Regan. He rewrote it and published it in my name. The paper appeared after the Geller tests. The late David Bohm did not like what I wrote and he wrote a rebuttal to it with Basil Hiley.

O'Regan and I were at Cambridge attending a meeting on psychic research sponsored, if I remember correctly, by Ted Bastin's group of new age physicists and computer scientists. Nobel prize physicist Brian Josephson [104] was there. After the meeting a sprightly Englishman walked up to me and introduced himself as Dennis Bardens.[105] He said:

"Dr. Sarfatti, may I take you to dinner?"

Fred Wolf was there and he suggested I go with Bardens. We had a good dinner of duck in cherry sauce at the Blue Boar Inn. After dinner, over brandy and cigars, Bardens leaned towards me with a conspiratorial wink and said:

"First, I want you to know that I am a Cabalist."

After a dramatic pause he continued in a more officious tone[106]:

---

[103] Jean Stein's father, Jules Stein (MCA) also was funding some New Age projects in the mid-seventies along with Laurance Rockefeller. One of Rockefeller's close friends Jean Lanier financed me at that time.

[104] Josephson spent two weeks with me in San Francisco in 1977. The visit was reported in The Chronicle. He spent a lot of time with Puthoff and Targ at SRI. I introduced Zukav to Josephson. Josephson believes that the superluminal quantum connection is necessary for life and consciousness to exist.

[105] Peter Maddock, who was also at the Cambridge meeting told me, years later, "Oh yes, Bardens, he was a part-time stringer for British Intelligence". Maddock resurfaced at the Tucson II meeting in April 1996 where he presented a paper on telepathy (abstract # 418)

[106] I had been at CIA Los Angeles Office near UCLA in 1963 when I was working at Ford Philco Aeronutronics in Newport Beach, California. This was also around the same time I first met Feynman at Cal Tech where he told me about his work in superfluid helium vortex formation that gave me the idea for my Ph.D. topic on rotational gauge fields in superfluid helium. A former CIA Chief of Station, Harold Chipman confirmed that I was part of a long-term operation in 1985. Chipman said he was running the SRI RV program behind the scenes without Hal Puthoff's knowledge. Chip mentioned Kit Green more than once. See also book "Future War" by Col. John Alexander" who wrote

"Dr. Sarfatti, it is my duty to inform you of a psychic war raging across the continents between the Soviet Union and your country and you are to be in the thick of it!" [107]

"The New Mental Battlefield" in Military Review, 1980 and directed Los Alamos Lab for Non Lethal Weapons of Mind Control. Col Alexander was on the board of advisors to Robert Bigelow's NIDS in Las Vegas. Bigelow is the new "Howard Hughes" with vast real estate wealth. Most of the real US UFO data analysis, using high level retired police and US military intelligence officers, was funded by Bigelow via his "National Institute of Discovery Sciences" (AKA NIDS). Unlike Joe Firmage, Bigelow still has money and power. There is an internet legend that James Jesus Angleton of the CIA was really concerned about flying saucers http://www.nexusmagazine.com/angleton.html. Stephen Schwartz who also generally discounts Chipman's allegations, says that story is a lie. Bigelow essentially, except for UFO website, closed down NIDS in the Spring of 2002 concentrating his waning post 911 hotel resources on Bigelow Aerospace in Las Vega working with NASA to bring ordinary tourists into near space orbit. Bigelow's agenda, however, still very much contact with alleged ET aliens possibly coming through Star Gates in the brane parallel universe (Jacques Valle's "Magonia" of Michael Murphy's "Greater Earth") less than a millimeter away from us across a thin Josephson junction hyperspace barrier. NIDS people had allegedly at least one encounter with an alien being on Bigelow's Utah ranch that would fit this scenario and is consistent with NIDS ex-physicist Eric Davis MUFON 2001 paper. Davis and Puthoff are still working together on UFO related physics allegedly with a USAF grant as this book goes to press.

[107] See Ron Mc Rae's "Mind Wars", also "Special Tasks" by Pavel and Anatoli Sudoplatov, Updated paperback edition with forward by Robert Conquest of the Hoover Institution at Stanford University p. 483.

JACK ANDERSON
1401 Sixteenth Street, N. W. Washington, D. C. 20036

Dear Jule –

Sorry I've been so long – here's a $100 to help you in your good works. More when available.

Boole research going well – conspiracy in state of panic.

Please write Williams – get copy of S21 Senate report if possible. Also ask ① when was psi discussed at NSC, ② was psi ever imported in bomb explosions aboard ships during Viet Nam (get specifics + reports if possible, and ③ address, if known, of Jack Ford (with DoD during Kennedy) + details of interest in psi.

Cordially,
Ron (McRae)
(HIA ~~special~~ special agent)

G-Strings and Flying Saucers by Jack Sarfatti

"Life is a beautiful thing as long as I hold the string." Sinatra

http://www.thepeaches.com/music/frank/IveGotTheWorldOnAString.txt

Enigma of the point particle

Herbert Frohlich[108] took me aside one day in 1967 outside the physics department office at Revelle College of the University of California, San Diego in La Jolla saying, "The idea of the point particle is what is wrong with physics." Nevertheless, scattering data on the electron shows it to be a point particle at very high energy. Similar deep inelastic scattering of electrons off the proton and neutron indirectly show three real point-like quarks with fractional electric charge inside each nucleon surrounded by a cloud of virtual particle-antiparticle source pairs and virtual force gluons. The problem with point particles is that they have infinite self-energy not even Feynman's genius was able to solve that enigma completely. Feynman was never satisfied with his "renormalization" trick for which he shared the Nobel Prize with Schwinger[109] and Tomonaga.

---

[108] Discoverer of the isotope effect that helped Bardeen, Cooper and Schrieffer formulate the theory of electrical superconductors as a quieting of the local randomness of the zero point fluctuations by the formation of coherent quantum superpositions of different numbers of nonlocally connected, or "real entangled" electron pairs . This violates, i.e., "spontaneously breaks" an internal "gauge symmetry" from hyperspace in the lowest energy "ground state". The same idea works for the virtual quanta inside the "quantum vacuum" of the system. This new coherent phase ordered quantum vacuum state with a smaller amount of random zero point fluctuations (ZPF) has a slightly lower total energy than the normal state with a larger amount of random zero point fluctuations. Symmetry means that something does not change when something else does change. P.W. Anderson was also at UCSD around this time where he gave his important 'More is different' talk. Anderson wrote: "I admired Frohlich's book on dielectric theory…I came to see the crucial importance of the concept of broken symmetry…Broken symmetry is the clearest instance of the process of emergence which lies behind 'More is different' …I developed the concept of the Goldstone boson…superconductivity, the Higgs phenomenon and the concept of Generalized Rigidity arose, the latter referring to the similarities between superfluidity superconductivity, and elastic rigidity of all sorts" from "A Career in Theoretical Physics", P W Anderson (World Scietific,1994)

[109] Julian Schwinger left staid Harvard for UCLA where he liked to drive Cadillac convertibles to the beaches. Feynman had a similar experience with me in 1963 down the Sunset Strip in my black Jaguar XK 150 convertible with white leather seats that I got at Hollywood Sports Cars.

Indeed, Feynman told me in his Cal Tech Office in 1968 that he thought what he did was no better than a "shell game" and that it was a "scandal" that no one was able to come up with a better idea. So Feynman agreed with Frohlich on this. They did not agree with Steven Weinberg's making a virtue of the necessity of renormalization, which is also very hard to understand and is probably not mathematically kosher. The amazing fact about it, however, is how well it works giving important numbers that agree with experiment to great accuracy.

Vibrating G-Strings[110]

As Saul-Paul explains below some people did begin to do better later on. The idea is that all the elementary particles are vibrating G-Strings. The great advantage to this is that the self-energy is no longer infinite and the theory based on Feynman's diagrams for point particles on world lines gets even simpler and more elegant. However, there is still a problem with the modern hyperspace string theory, which is that the energies are really much too high. When you combine the basic numbers of gravity with the basic numbers of quantum theory you get a G-String length of 10^{-33} cm, which is 20 powers of ten smaller than the proton at 10^{-13} cm[111] should be. The electron is actually larger at around 4×10^{-11} cm, i.e. 22 powers of ten bigger than the G-String. On the one hand, the string length is so tiny that it would seem to explain why the electron looks like a point smaller than at least 10^{-16} cm in scattering data. On the other hand, basic quantum theory says that the electron should really be quite large at 10^{-11} cm called the "Compton wavelength".[112] So we have a paradox that I solve for the first time in this book with "The G-String Theory". To be more precise I should use the symbol G* rather than G that is the usual symbol for Newton's constant of gravity measured in the laboratory.[113] The G* theory was first introduced almost thirty years ago by Nobel Laureate physicist Abdus Salam. I worked on it with him in Trieste, Italy in 1973. In this book, I apply that idea in a new way. The idea is that G is not a constant but is a variable G* getting very large at small distances. Once we have that idea, then the electron is actually a vibrating G-String of length 10^{-11} cm, but because of the strong short range warping of space at that scale it looks like a point particle when one probes it with something called the "spacelike virtual photon". This effect is crucial for an understanding of the way a flying saucer might fly. We shall see later that there is a "spacetime

---

[110] Never underestimate the power of a vibrating G-string!

[111] Called a "fermi".

[112] $\hbar/mc$ = 3.861592 x 10^{-11} cm for the electron for those who know some physics.

[113] G = 6.670(4) x 10^{-8} cm^{3}sec^{-2}grams^{-1}

stiffness coefficient" that is 1 fermi per 4 billion metric tons.[114] Remember, the proton is about 1 fermi across. In contrast, the neutral hydrogen atom of a single proton and a single electron is about 10^{-8} cm across in its lowest energy ground state. That is 100,000 times larger. Spacetime is simply much too stiff to bend with an electromagnetic control field if G is really an absolute constant. [115] This means it is impossible to make a flying saucer machine that will generate its own timelike geodesic warp drive in which every thing inside is in weightless LIF free float with no g-forces, no need for pressure suits. In contrast to an LNIF observer fixed to the surface of the Earth, or even in a Phantom Fighter with afterburners on, the saucer appears to make a hairpin turn at Mach 10![116]

---

[114] From Feynman's Cal Tech Lectures on quantum gravity.

[115] This is why Eric Davis's claim at 2001 MUFON that Hal Puthoff's theory explains flying saucer flight physics is wrong. Eric has not responded to my challenge for a detailed justification - a strange silence.

[116] I do not think the physics in the new book "The Hunt for Zero Point" by Nick Cook of Jane's Defense Weekly on this is on target.

My World Line by Jack Sarfatti

The Tibetan Connection

I relate a very uncanny Jungian synchronicity. I was attending a meeting of the AAAS at the Hilton Hotel in 1980.[117] A man walked up to me noticed my name tag "Sarfatti" and said "Are you related to Margherita Sarfatti?" "Yes, distantly." I replied. "I have traced your family tree back a thousand years to Rashi de Troyes. My name is David Padwa." Padwa told me a little about the great French Rabbi. He claimed that one of Rashi's daughters went to Spain and that her descendants formed our Italian branch after the expulsion of the Ladino-speaking Spanish Jews in 1492. He said we were distant cousins. He gave me his card. He was from Santa Fe, New Mexico. I mentioned the incident to my buddy, Beat Poet, Gregory Corso. [118]

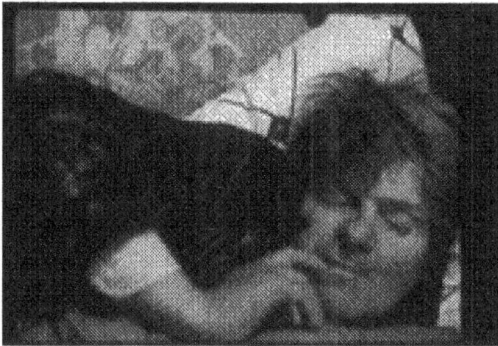

Gregory Corso sleeping in Jack's North Beach Pad. Photo by David Gladstone

---

[117] I also met Marcello Truzzi within minutes of meeting Padwa shortly before John Wheeler attacked parapsychology in this famous meeting of the AAAS.

[118] Photo by David Gladstone; the late Beat Poet Gregory Corso sleeping in my North Beach Flat 1980's. I took Gregory to a party at Lawry Chickering's in Sara Fernandez's Mercedes. Sara ran the Fritz Jewett Foundation. Herb Gold, Sandra Mossbacher and William F. Buckley, Jr. were there and, of course, "Gwegowy" made a scene! This was when Buckley asked me how to attack Carl Sagan on Nuclear Winter as we sat in the back seat of his black limo right before he went to TV station to debate Sagan.

Corso[119] said.

"Oh yeah, Padwa! I knew him in the Village. He's a smart guy who made a lot of money from Xerox. He went with Ram Dass to India. He's mentioned in the book "Be Here Now" as the rich American in the Land Rover. He's a real heavy with the Dalai Lama. He brought Tibetan Tulkus to America."

About a month went by. Dave Massetti, editor of North Beach Magazine back then hands me a letter from Padwa. I open it. Padwa writes that a strange thing happened to him right after he met me. He had been looking for the autobiography of George Gamow's "My World Line" for six months. It was out of print. Mathematician Stan Ulam had told Padwa to read it. Padwa leaves me at the Hilton after the conversation about Rashi wanders over to North Beach and discovers "My World Line" on a shelf in Discovery Bookstore. He then goes to Ferlinghetti's City Light's Bookstore and sees a copy of North Beach Magazine with my picture on the cover. In the picture I am holding a copy of Gamow's "My World Line" in front of a large poster of Einstein. Here is the picture by Bob Jones, which I finally found from John Shaw in North Beach on Jan 11 1997. It is from Jan 1980 issue of North Beach Magazine.

---

[119] David Padwa visited Gregory Corso as he lay dying.

My World Line 1979 Cover of North Beach Magazine by Bob Jones

I reproduce the following information for the record.[120]
http://history.acusd.edu/gen/filmnotes/cradlewillrock3.html
http://www.mgm.com/teawithmussolini/

The Cradle Will Rock - Tea With Sarfatti

SARFATTI-GRASSINI, MARGHERITA...A mistress, confidante, and biographer of Mussolini, Sarfatti was a highly influential figure in the cultural and artistic policies of the Fascist regime...from a comfortable Venetian-Jewish family...A woman of acute intelligence and sophistication, while still in her teens she became a militant activist in the Italian Socialist Party...and in the feminist movement along with such major Socialist women as Angelica Balabanoff, Anna Kuliscioff, and Clara Zetkin...She...married a Socialist lawyer, Cesare Sarfatti, and both became close friends and supporters of Mussolini...Their son Roberto (1900-18) who was killed in the war, was later an object of Fascist veneration...In the years during the struggle for power Sarfatti and Mussolini became intimate friends, and she exercised an increasing influence on him...she was dubbed the 'dictator of the figurative arts.' Sarfatti was one of the most avid admirers of twentieth-century modernism...her political influence enabled her to...sponsor painters, sculptors and architects...In 1925-26 Sarfatti was instrumental in founding the Novecento art group in Milan...by 1934 their intimate relationship had ended...The final break came in 1938 with the passage of the Fascist anti-Semitic laws...In 1939 she left Italy on a passport provided for her on Mussolini's instructions...

- Historical Dictionary of Fascist Italy, Phillip V. Cannistraro (Greenwood) [121]

Mussolini's granddaughter is active in politics. She is the niece of Sophia Loren.

---

[120] The Tim Robbins film, "The Cradle Will Rock" features Susan Sarandon as Margherita Sarfatti.
[121] More details are in "Il Duce's Other Woman" by the same author with Brian R. Sullivan. The Sarfattis and the Grassinis were also blood cousins and there is an obvious strong genetic similarity to my father's sisters Regina and Victoria Sarfatti and I to her son, Amedeo.

אלה אנשי חיל
נפלו גבורים בתוך המלחמה
שמם וזכרם לדר דר׳

GLI EBREI VENEZIANI
CADVTI IN GVERRA PER LA PATRIA
LA COMVNITÀ
CON AMORE CON ORGOGLIO RICORDA

ABOAF UMBERTO
ANCONA PAOLO
BORALEVI GIORGIO
FOA' DAVIDE GVIDO
GRVNWALD BENIAMINO
LEVI BIANCHINI ANGELO
LEVI MINZI GVIDO
LEVI MORENOS ALBERTO
LEVIS GIVSEPPE ERNESTO
NACAMVLLI MARIO
NAVARRO ABRAMO
PADOA ALDO

PARDO GIORGIO
POLACCO ABRAMO
POLACCO SANSONE
SARFATTI ROBERTO
SEGRE IPPOLITO
SOAVE AMEDEO
SOAVE ATTILIO
SONINO OLIVIERO
STECHER BRVNO
TODESCO MARIO
VIVANTE FERRVCCIO
FINZI RVGGERO

MCMXV · MCMXX

The Pope's Jew [122]

I was at the Rev. Moon Unity of Science Conference on the Absolute at the Fairmont Hotel in 1980. I met Yuval Ne-eman [14] and Max Jammer from Israel. Ne-eman was quite interested in talking about the significance of the name "Sarfatti" to Jewish History. Ne-eman was, of course, talking of Rashi - the first Sarfatti [15]. Ne-eman has published an important paper on the fiber bundle mathematics of the quantum connection beyond space-time.

Samuel Sarfatti was the personal physician to Pope Julius the Second. Samuel was a friend of Michelangelo. He taught anatomy to Michelangelo. Sarfatti used his influence with the Pope to get Michelangelo the commission to paint the ceiling of the Sistine Chapel. Michelangelo shows God reaching backwards to Adam. This is the perfect symbol for the root idea of my new physics that the future creates the past. God evolved from man in our future uses time travel to create the universe and man in what physicists now call a "globally self-consistent loop in time".

Part of my new physics message is that synchronicity can be an effect of contact with advanced intelligence able to manipulate the quantum connection in what John Lilly called "cosmic coincidence control". This is a dangerous idea, but it may be true. It is a scientifically testable idea. Quantum devices can be built based upon it if it is true.

I do not mean to imply that every paranoid fantasy by mentally disturbed people should be accepted on face value - but I suggest, some of the "voices" people hear might be from elsewhere. Each case must be studied individually in a scientific objective way.

---

[122] Margherita Sarfatti's father Amedeo Grassini was a long time close personal friend and financier to Pope Pius X. He was called "The Pope's Jew". Margherita was called "The Uncrowned Queen of Italy" in the 1920's. The Grassinis and the Sarfattis were related both by blood and marriages. Margherita, in her 1926 official Fascist biography "Dux", wrote of Mussolini's alleged encounter with a Gypsy as child who foretold his sinister and tragic date with destiny. This, admittedly totalitarian propaganda piece, sounds like a Verdi Opera, very Italian melodramatic, I know. However, it is similar to Winston Churchill's childhood experience as well as to my 1953 experiences. These precognitive prophetic events of the Destiny Matrix are common in, indeed at the core of, the folklore of, human history in the stories of Moses, Joseph, David, Alexander, Caesar, Jesus, Mohammed et-al.

I think the extreme relativism of New Age thinking is wrong. True, there are many complementary points of view equally valid within their proper context, but the whole idea of physics is that there are objective absolute "invariant" truths.

The Occult Third Reich

It was 1978. I had recently written a black comedy called "Hitler's Last Weapon"

http://qedcorp.com/book/psi/hitweapon.html

It was about a New Age Guru who is the reincarnation of Hitler and becomes the first psychic dictator of the United States. This was not so far fetched then. My old boy Cornell chum, Lee Myers, had paid Steve Hill[123] to make a twenty minute radio program of the script with British actor Eric Bauersfeld narrating. The program became a minor cult classic on public radio and is still played late at night.

Hitler's "Private Buffoon - a light-hearted loon"[124]

http://www.tmbhs.com/tmbhs/movies/theproducers/theproducers.asp

One day I was sitting on the terrace of the Savoy Tivoli on Grant Avenue in North Beach. I believe, but I am not certain, that Leila Minturn Dwight introduced me to a handsome young man of about nineteen that she said was her distant cousin from Munich. The young man, Eric (or Egon?) Hanfstaengl was here for the summer and soon became a regular part of my circle of Caffe Trieste cronies. I'm not sure I got the first name right. I did not pay too much attention to him as I was more interested in meeting women and I have avoided parental roles. I did invite him to a party given by bon vivant Norwood Pratt. The young Hanfstaengl came in lederhosen and sang Tyrolean songs. In the course of casual conversations he had indicated that his grandfather Ernst Franz Sedgwick-Hanfstaengl - (a.k.a. Putzi) had played an important role in history. He was never specific. I did not realize then that his grandfather had been the Victor Borge of The Third Reich personally beloved by Adolph Hitler. I did not know then that the young man's father sat on Hitler's lap many times and called him "Uncle Dolph". I did not know that the young Hanfstaengl's grandfather was depicted in Syberberg's "Our Hitler" which I had seen. The end of summer came. Young

---

[123] Hearts of Space Radio, Pacifica

[124] "If you listen to popular rumor." The Yeomen of the Guard, G & S, also Mel Brooks, "The Producers"

Hanfstaengl's wife-to-be came, she was the daughter of the publisher of Der Stern. The two innocent children went back to Munich and I hope are living happily ever after.

Putzi's book "Unheard Witness" is dedicated to his friend Oswald Spengler. Like many of the Sedgwicks, Putzi had a gift for writing.[125] Here are a few tantalizing short literary bytes from his table of contents:

"My schooldays with Himmler's father - Sedgwick, Heine and Hanfstaengl forbears - Harvard and Theodore Roosevelt...The American military attaché speaks of Hitler-Introduction to an agitator...Introducing Hitler to society...Wagner on an upright piano...Plan for a putsch... Hitler's attempted suicide in 1923..."

Hitler's 1923 suicide attempt foiled at last moment!

Here hangs a tale in which the inaction of Putzi's wife, Fr. Helene Niemeyer Hanfstaengl, could have prevented World War Two and the development of the atomic bomb by my Cornell physics professors who were at Los Alamos. In fact, what happened in this universe, was that she stopped Hitler from shooting himself in the head, at the very last moment, by knocking the gun out of his hand just as the police were about to enter.

If the many material "brane" worlds of Super Cosmos[126] are correct, then, at that dramatic moment the universe split into two parallel universes. In the universe next door the Holocaust never happened and neither did the State of Israel. Carlo Suares speaking from the occult cabalistic perspective said that Hitler was God's Instrument for the Restoration of Israel. I do not think it was worth the price. Putzi continued:

"The log-chest in the corner of the fireplace in my library is still covered with the traveling rug I lent to Hitler when he was a prisoner in Landsberg...It

---

[125] Catherine Maria Sedgwick, "A New England Tale", first American female novelist, daughter of Judge Theodore Sedgwick, first Speaker of the House after the American Revolution, Alexander Hamilton's mentor. Catherine had a literary salon that included Herman Melville, Nathaniel Hawthorne, William Cullen Bryant et-al. Ellery Sedgwick, Henry Dwight's brother, was the great editor of The Atlantic from 1909 to 1938. Henry Dwight Sedgwick's nickname was "Babbo". All the Sedgwicks made it to Harvard's most exclusive Porcellian Club. FDR did not get in! I don't recall off hand if Putzi was a member?

[126] Hawking's "The Universe In a Nutshell".

was to my house in Munich…that he came for his first meal after release from jail and where, nearly a decade later, he celebrated with Eva Braun the year of his triumph. Mine was the first Munich family of standing into which he was introduced in the days of his insignificance.

…I tried to impregnate him with some of the norms and ideas of civilized existence, only to be thwarted by the ignorant fanatics who were his closest cronies. I fought a running and losing battle against Rosenberg and his hazy race mystique…People have said I was Hitler's court jester. Certainly I used to tell him jokes, but only to get him into the sort of mood in which I hoped he would see reason. I was the only man who could hammer out Tristan and the Meistersinger to his satisfaction…My mother was born a Sedgwick-Heine. My maternal grandmother came from the well-known New England family and was a cousin of the General John Sedgwick[127] who fell at Spotsylvania Court House in the Civil War and whose statue stands at West Point. My grandfather was another Civil War general, William Heine. In the funeral cortege of Abraham Lincoln he was one of the Generals who carried the coffin. My mother…could remember Lincoln's funeral clearly…The Hanfstaengls…For three generations…were privy councilors to the Dukes of Saxe-Coburg-Gotha…in 1905 I was sent to Harvard…I made friends with…T.S. Eliot, Walter Lippmann, Hendrik von Loon, Hans von Kaltenborn, Robert Benchley and John Reed…President Theodore Roosevelt…had heard of my prowess through his son and invited me to Washington in the winter of 1908…I…took over the Hanfstaengl branch on Fifth Avenue. It was a delightful combination of business and pleasure. The famous names who visited me were…Pierpont Morgan, Toscanini, Henry Ford, Caruso, Santos-Dumont, Charlie Chaplin, Paderewski…I took most of my meals at the Harvard Club , where I made friends with the young Franklin D. Roosevelt." [128]

Putzi finally fled Hitler in 1938 in fear of Goebbels's jealousy of Hitler's fondness for him. Putzi's distant cousin FDR was embarrassed by Putzi's presence in Washington and sent him to an obscure Army post in Texas. Sam Rayburn pressured FDR into accepting the Pentagon design for the new war office as part of the Putzi deal. FDR wanted another design.

---

[127] General Sedgwick was a popular charismatic leader who Lincoln was going to appoint as head of the Union Army of the Potomac that fell to US Grant. Sedgwick is alleged to have said "Don't worry boys, those Rebs can't hit the side of a barn" moments before a fatal wound to the head by a Confederate sniper.

[128] I am reminded of September 1960 newly arrived at Brandeis University as a National Defense Fellow. I was on my way to the home of President Abraham Sachar. A frail old lady came up behind me and asked if I would help her up the small hill to Sachar's house. It was Eleanor Roosevelt. She entered the party holding my arm. I also met Isaac Bashevis Singer sitting on a bench on campus. I had already read some of his novels.

Francis Ford Coppola distributed Syberberg's film "Our Hitler" Putzi is a major character in that film. I did not realize the connection until years later. I saw "Our Hitler" around the same time that I met Putzi's grandson in North Beach.

Fascism, The Dark Side of Bohemia by Stephen Schwartz

"Fascism was a mass movement of the disfranchised and disgruntled middle classes in Europe, whose position was threatened by social instability: that of the Italian regime in the face of proletarian revolution after World War I, in the case of Mussolini's movement, and by the economic hardships in Germany following the collapse of the empire in the same war, in the case of Hitler.

Konrad Heiden, an anti-Stalin leftist and the first great biographer of Hitler, wrote the earliest really useful social analysis of fascism. He described its beginning as a movement of "armed intellectuals," then becoming a movement of the "armed bohemia". Trotsky carried this analysis forward, but being blunter, described fascism as "the panicked radicalism of the ruined and crazed petit bourgeoisie," a movement composed of "human dust"

"Human dust." Let us examine what this term and "armed bohemia" mean.

Beginning in the 19th century, with the exponential growth of the great European cities, interesting sociological developments are noticeable.

Prior to then, society was composed of fairly stable classes: the aristocracy and church, the commercial bourgeoisie, the trade and craft productive class, the peasantry, and, weak but noticeable, the preindustrial wage-working class. Trade and craft workers were paid for their products; wage workers for their time.

With the new rise of the cities, the decline of the old ruling and trade/craft classes, the growth of the commercial and protoindustrial bourgeoisie, ruination of the peasantry, and the continuing development of the protoindustrial working class are all visible. But other sociological phenomena are also perceptible. A new caste or class emerges—the inferior section of the petit bourgeoisie, consisting of:

- aspiring state office holders;

- parasitical entrepreneurs unable to establish a firm place in the new economy;

- "overeducated" but mediocre placeseekers in the liberal professions, including academia and what we now call media;

73

- ruined trade and craft workers and peasants, devastated by the new economy;

- those who migrated from the country to the city and became demoralized, along with other declassed elements, by the pace and difficulty of urban life, to a point of unemployablity, alcoholism, criminality, and mental incompetence.

This vast array of unstable elements formed the bohemian class in the cities.

They seldom had fixed jobs. They were filled with resentment at their failure to establish themselves in society. Some of them were also enraged at the social and economic changes that had deprived them of their place in society. But most of them just hated the position they were in; they could not reach the point of analyzing their predicament.

So they turned to conspiratorial explanations.

Sound familiar? No stable profession or job? Without a hope for the future. Filled with anger? The same group of disgruntled nobodies.

Human dust.

It is the special characteristic of this class—the urban bohemia—that it cannot think straight. It always seeks a scapegoat for its failures, and its thinking is always conspiratorial.

- The declassed bohemia of 19th century Catholic Europe blamed its misfortunes on a Jewish-Masonic conspiracy.

- The declassed bohemia of post-1918 Italy blamed its misfortunes on the "antipatriotic" socialists.

- The declassed bohemia of post-1918 Germany blamed its misfortunes on a Jewish-Masonic-Bolshevik conspiracy.

- The declassed elements of Spain in the 1930s blamed its misfortunes on a Jewish-Masonic-Communist-anarchist conspiracy.

- The declassed bohemia of collapsing Serboslavia blamed its misfortunes on a Catholic-Muslim-NATO conspiracy.

The sociological profile is exact: aspirants to employment in the raggedy-assed Tito bureaucracy, wannabe entrepreneurs who become Belgrade mafiosi; "overeducated" place seekers in the liberal professions, including academia and what we now call media—Yugoslavia had more unemployed film critics than any country in history, and the Serbian press was and is as bad as it was a century ago; ruined trade and craft workers and peasants - the Balkans had a hell of a lot of these; those demoralized by the pace and difficulty of urban life to a point of unemployability, alcoholism, or mental incompetence—the Balkans were also full of these.

There is even a declassed "bohemian" stratum in the Islamic countries—indeed, it is enormous. In some countries they are known as the "teahouse class." [129] This class consists of those who fail to gain state employment, or to start businesses, or who have too much education and no prospects, or whose peasant holdings or trade and craftwork have collapsed, or who cannot contend with urban life. And of course they blame their misfortunes on the Jews and America.

The declassed bohemia of North Beach blames its misfortunes on the CIA-Nazi conspiracy, and occasionally takes up the cudgels for Serboslav or Islamic conspiratorialism. The teahouse class in Cairo = the coffeehouse class in San Francisco.

The situation has been much aggravated by the rise of the universities as, essentially, glamorous diploma mills - the mass fabrication of doctorates, etc.—and by the incredible over expansion of the state bureaucracies worldwide after 1945. There were millions of these types in Imperial Germany and Austria-Hungary. They had no real education, no serious profession, but they got themselves parasitical posts in a ridiculously overgrown state administration. Look at Hitler's father, the petty customs collector. These jobs don't even exist today. No stable profession, but marginal jobs for marginal enterprises that bloom in periods of speculative accumulation and then collapse. [130] Millions of

---

[129] A pool of potential recruits for Osama bin Laden's al Qaeda.

[130] E.g. the rise and fall of vaporware Dot.Coms in San Francisco. - Sarfatti comment
I know from direct experience in San Francisco that the Dot.Com bubble was inflated by style over substance, by New Age Cargo Cult psycobabble of gurus, flim flammers and hucksters that together with "crime in the suites" at ENRON, Arthur Anderson, World Com et-al has seriously weakened America and the world economy at the worst possible moment with Terror at The Gates. To her credit, Marshall Naify's wife, Lily, warned me that the Dot.Coms of Joe Firmage and other Gen X Carpetbaggers who descended on Silicon Valley and SOMA like locusts, would completely collapse months before they did. Unfortunately, I did not listen to her sound advice.

such people were thrown out of work in Germany under Versailles. They marched right into the Nazi party.

The Bay Area is loaded with overeducated place seekers; its leftism is really fascism.

All of these types are filled with resentment. Resentment that they do not have better jobs, that they are not recognized for their genius, etc. The junior version is John W. Lindh.

A century ago the socialist labor movement offered an alternative to all this: the industrial workers had a productive position in society, and they embraced a rational, nonconspiratorial radicalism. But that class no longer exists, and as Franz Borkenau noted in the 1930s, German Communism turned from organizing the workers to competing with the Nazis for votes from the armed bohemia. That was the end of that, and it explains why the Commies of today have become fascists—defending Serbia, for example.

I fear the real Orwell/PKDick future will consist of a handful of serious professionals—those able to establish themselves in "letters and science"—surrounded by a vast, limitless human desert of fakes, poseurs, conspiratorialists…

UFOmania is another example of this phenomenon: the declassed Americans and Europeans baffled and besieged by the world look for yet another conspiratorial escape hatch. Faced with too much science for them to understand, they flee into pseudoscience.

Not a pretty picture. The conspiratorialist poseurs do more to establish fascism in America with every cappuccino they drink than an army of a million Byelorussian exiles, bankrolled by General Motors, General Mills, and General Foods, and commanded by General MacArthur, could ever have accomplished."[131]

---

[131] January 5, 2002 email from Stephen Schwartz to Jack Sarfatti "For myself, I see the anti-banking rhetoric and anti-Jewish rhetoric (I don't use the term anti-Semitic because Arabs are also Semites) as pretty much the same. I am not sympathetic to the anti-IMF and anti-global protests. Marx, Engels, Lenin, Trotsky, were all globalists—Stalin, like Mussolini, Hitler, and the Japanese imperialist, was not. The historical left always opposed protectionism, and turn in that direction shows that the moral underpinning of the traditional left is gone. They are collapsing into fascism." Email-from Schwartz, June 18, 2002 on "Ezra Pound". Note it was Stephen Schwartz who defends Heisenberg from Nazi smear. Meanwhile, I don't care if Hick Nerdbutt finds a cure for cancer. No

"Ciao! Manhattan!"

My connection to the Sedgwick karma thickened because two years later in 1981, Leila Minturn Dwight introduced me to her aunt, Suky Sedgwick, who was Edie Sedgwick's baby sister. I fell in love with Suky's magical qualities. I did not know about Edie, nor was I consciously aware of the connection to young Hanfstaengl.

The Sedgwicks seem to have an Orphic connection to Hades as might have been imagined by Edgar Allen Poe. John P. Marquand, Jr., a Sedgwick himself writes:

"Have you ever seen the old graveyard up there in Stockbridge? In one corner is the family's burial place: it's called the Sedgwick Pie …In the center Judge Theodore Sedgwick is buried…his tombstone, a high rising obelisk…The legend is that on Judgment Day when they arise and face the Judge, they will

---

scientific or other intellectual work can justify Holocaust revisionism. I consider Pound the greatest English language poet of the 20th century but his fascist speeches remain wretched and despicable, and he deserved to have been imprisoned for them, though he was not since he played crazy, nut Hamsun was among the greatest prose writers of modern times yet he was sanctioned by the Norwegian public as a collaborator and boycotted after world war 2. Many of my friends consider Celine a great novelist. I don't. I think he should have been executed as a Nazi collaborator. On the other hand, there are two pitfalls to be wary of. One is that leftists who exaggerate the culpability of fascist and Nazis never want to discuss the guilt of criminals like Pablo Neruda, who was a conspirator in the murder of Trotsky. The other is the temptation to exaggerate. I consider Pound to be guilty of a major crime: he made horrific propaganda speeches during the war. Celine wrote pamphlets advocating the extermination of the Jews, another major crime. By contrast, Heidegger has been pilloried by the left when his collaboration was minor and consisted of no more than a couple of academic lectures supporting the Nazi regime at its beginnings, the firing of Jewish professors. I consider Heidegger to have been unwise but not culpable in a serious way. I consider Bohr, Oppenheimer, Morrison, and Bohm much worse than Heidegger. Bohm deserves recognition for the genius of his physics but he was a spy for the Russians and probably deserved to be locked up for it. The art of politics is the art of making distinctions, not confusing them.

Holocaust revisionism is unacceptable in any form."

Stephen Schwartz e-mail June 16, 2002 responding to Nick Herbert's continuing surprising disturbing defense of "historian" David Irving. Nicks likes to use the cyber-alias "Dr. Jabir".

have to see no one but Sedgwicks…He was a political ally of Alexander Hamilton and George Washington…Minturn had always been very much involved in the traditions of the Pie…Kennedy knew all about the Sedgwick Pie, and Minturn wondered after having watched the…Kennedy funeral - the casket on the horse drawn cart…if perhaps Mrs. Kennedy hadn't 'borrowed' the idea from us…He stocked up on simple coffins…got into them and 'tested' them."…Pp.3-9 EDIE by George Plimpton and Jean Stein.(Knopf, 1982)

Jack Sarfatti and Suky Sedgwick mid-80's

Suky was charming brilliant and sexy. She loved to give dinner parties to the music of Ella Fitzgerald [9] and Louis Armstrong. Suky played classical piano and was a great Italian cook. Like the Henry Jamesian American Princess abroad that she was, she had married an Italian count and was fluent in Italian as well as French. At the time I did not know that I could also have claimed Italian Nobility because of the Sarfatti Crest on a wall of the Synagogue in Venice given by the Doge.

STEMMI

DI FAMIGLIE ISRAELITICHE

SARFATTI

Sarfatti Crest, Venice, Italy

Suky's bedroom contained a leather-bound set of the autographed works of Rudyard Kipling that she inherited from her De Forest grandmother.[132] Her grandmother nursed Kipling to health after a severe illness at their Long Island estate. British Parliament opened with "God Save Kipling." and they sent a letter of thanks to Suky's grandparents.[133]

David Weisman's and John Palmer's film "Ciao! Manhattan" stars Edie. The plot involves a UFO contactee in a relation to Edie which strangely precognizes my future relation to Suky. I did not see the film until my romance with Suky was over—for then, at least...

---

[132] Same De Forests that Humphrey Bogart also came from. Suky knew Lauren Bacall. Suky's grandfather built Noyes Lodge at Cornell near the girl's dorms where I used to hang out. Julia Noyes was Suky's grandmother. Alice Delano De Forest was Suky's mother, same D as FDR. Suky was named for Susanna Shaw Minturn whose brother Robert Gould Shaw, her great great paternal uncle, led the 54th Massachusetts Black Regiment in the Civil War. The story of the 54th is in the film "Glory". General John Sedgwick's statue stands at West Point. John Sedgwick, killed at Spotsylvania, was Lincoln's favorite. John was replaced by U.S. Grant. FDR's Groton school master Endicott Peabody was part of the large Sedgwick Clan.

[133] Henry De Forest was a close friend of both of John Foster Dulles and Allen Dulles.

Suky's grandfather Henry Dwight Sedgwick wrote a prophetic essay "House of Sorrow" in 1908 in which he says that the future creates the past. His imagery is almost identical with that of Sir Fred Hoyle's in his book "The Intelligent Universe." I believed back then in the 1980's that the quantum force of destiny, or the Ghost of Henry Dwight Sedgwick, had guided me to Suky. I would not have known about Henry Dwight Sedgwick's idea that the future causes the past were it not for Suky.[134]

---

[134] Kyra is Suky's cousin. Kyra is married to Kevin Bacon.

"More is Different" (P.W. Anderson) by Jack Sarfatti

I was at UCSD in 1967[135] where and when P.W. Anderson gave his first "More is Different" talk soon after his discovery of spontaneous broken symmetry. I had already written my broken symmetry paper on the Goldstone Theorem and the Jahn-Teller Effect with Marshall Stoneham at UKAERE in Proceedings of the Physical Society of London cited in "Resource Letter on Symmetry in Physics" of the American Institute of Physics. I was also starting to teach as an Assistant Professor of Physics at San Diego State University.

Jack Sarfatti at UK Atomic Energy Research Establishment

Theoretical Physics Division. 1966
Outside of Ridgeway House, Didcot, Berks

I told Stoneham about Goldstone zero modes and Stoneham told me about the Jahn-Teller Effect in crystals with the Mexican Sombrero Potential. We then did "1 + 1 = 2" a year before the time P W Anderson is talking about below. Stoneham replaced dislocation physicist Ron Bullough later as Director of

---

[135] With the Benford twins, Herbie Bernstein, Harry Yesian, George Chapline Jr. in his black Shelby Cobra down from Cal Tech frequently, James Edgar Felten, Arty Wolfe,, Bob Gould in the shadows of Keith Brueckner and Geof Burbidge, Bernd Matthias and their visitors like Fred Hoyle, Ed Teller, John Archibald Wheeler, Herbert Frohlich et-al.

Theory Group of Her Majesty's Nuclear Establishment. It was Ron who invited me to Harwell. I would also go over to Culham Plasma Lab and to Oxford. I learned about Goldstone stuff directly from spying on Kibble and Abdus Salam in their offices at Imperial College of Science and Technology in London where I would go frequently from Harwell. I also met Bohm at Birkbeck then. Salam was a bit contemptuous of Bohm's philosophical interests. Salam later invited me to IAEA/UNESCO Trieste, Italy in 1973 at a moment of high weirdness with SRI & Uri Geller as told by Martin Gardner in "Magic and Paraphysics" ("Science, Good, Bad and Bogus")

This new book "More is Different" (MID) Princeton, 2001 just came out. It is relevant to my use of broken symmetry in the "superconducting" physical vacuum, of off mass shell virtual particle antiparticle pairs and virtual boson condensates, to explain the zero cosmological constant, the anomalous acceleration in the expanding universe, Sakharov's idea[136] that derandomizing the zero point fluctuations (ZPF) gives warped spacetime as a classical limit, elementary particles as Wheeler vacuum geons

- Mass without mass

- Charge without charge

- Spin without spin

In Salam's strong short range G*ravity with $L_p^* \gg 10^\wedge\text{-33}$ cm, universal Regge trajectories of the lepto-quarks, charge to mass ratio and why vibrating lepto-quark strings look like point particles in scattering data.

From Nobel Laureate P.W. Anderson's article in that collection of essays:
"In the spring of 1967...I spent a pleasant month at La Jolla with...some old friends from Bell [Labs]...1967 was a temporary maximum of arrogance among the particle physics establishment, riding high in government advisory circles

---

[136] Essentially misunderstood, in my opinion, by Haisch, Rueda and Puthoff in their attempts to derive inertia and gravity from the purely uncontrollably locally random electromagnetic zero point fluctuations. The latter is only a small piece of the "normal" virtual fluid component of the quantum vacuum of the unified field. You need the coherent virtual superfluid to get both inertia and gravity..

(this was the heyday of JASON[137] and…RAND[138]…I used my newly generalized idea of broken symmetry to illustrate the processes by which this kind of autonomy arises…in which…truly novel properties and concepts…emerge from a simpler substrate. Broken gauge symmetry, which is exhibited by superconductors…Wigner's error…the 'superselection rules'…the idea that it would be meaningless to have coherence between states with different numbers of particles, i.e. that ground states are always discrete eigenstates of the symmetries of the problem. This misses the possibility…that macroscopic systems can have zero Goldstone modes, which restore the broken symmetry…, this concept I had encountered in my theory of antiferromagnetism in 1952…1967…this was before the renormalization group…in statistical physics and more or less simultaneously with electro-weak theory's publication…Even farther in the future…1975 was to come the topological theory of order parameter defects such as vortices, flux lines, domain boundaries…in terms of broken symmetry…all of the properties…crystal structure, metallicity, macroscopic quantum coherence, elasticity…nuclei containing only ~ 100 nucleons exhibit unmistakable evidence of…shape and superfluidity…"[139]

---

[137] Top National Defense Group of physicists that Keith Brueckner and other UCSD faculty were members of. I was working for Brueckner at the time. JASON has run into a dispute with Rumsfeld's DOD in the Bush Administration over who should be in it.

[138] USAF Think Tank in Santa Monica, California.

[139] The New York Times recently ran an article about this Dec. 4, 2001 "Challenging Particle Physics" http://www.nytimes.com/2001/12/04/science/physical/04SQUA.html. I show explicitly how classical spacetime emerges out of the "fire" of the quantum foam as a Josephson phase modulation effect in which the "world crystal" (Hagen Kleinert), like "ice", freezes out.

Tilting at Windmills; Jack Sarfatti's Impossible Dream? By Jagdish Mann

Jagdish Mann, Saul-Paul Sirag, Jack Sarfatti, 2001

Jack Sarfatti and I met in the summer of 1970 when he was teaching physics at San Diego State University and I was a student. We have been friends ever since. Temperamentally, we are as different as two men can be. He is an iconoclast, intense, full of nervous energy, impulsive, impatient. I, on the other hand, am a traditionalist, easy going, almost fatalistic, languid and patient to the point of laziness. Yet the mysterious and inexplicable force of friendship has favored us for more than thirty years. It has not only endured but also thrived on our differences. So, although I am not a physicist, I feel qualified to offer an introduction to this man.

How does one take a measure of another human being? The Good Book simply advises us not to do it, "Judge not lest ye be judged". This is good advice for all time. An individual human being is a dynamic bundle of contradictions and cannot be accurately categorized. A judgment is static by its nature, a thing frozen in the moment. To judge a man is to reduce him to a snapshot; turn him into a thing. "It is to change him from Thou to an It", said Martin Buber. Besides, you never get the whole story. Judgment is always a limiting act. You end up with a handful of water when you think you have caught a wave. This introduction is, nevertheless, a judgment of sorts. I would have liked to restrict it to his ideas alone but Sarfatti is not the kind of guy you can keep out. All through his theories, he keeps popping up like an impish Jack-Out-Of-The-Box. It is not so much that he wants to be judged, he wants to be measured, scrutinized. It is a brave stance.

It must be admitted that I speak with the bias of a friend. What may be to me 'perseverance' may be merely 'pig-headed stubbornness' to someone else; what I may look at as 'large dreams' may be called 'megalomania' by others; one man's visionary is often another man's fool. Only Jack Sarfatti's peers in physics, and of course time, the final judge of all things, *including his peers,* will tell if his theory is correct. I can attest to his single-minded dedication to his subject and to the sincerity with which he believes in himself. I am in a unique place to say this. He is writ large both in his faults and his virtues. But even during the worst times of his personal life, he has never abandoned physics. This kind of dedication is a necessary, if not sufficient, condition of creative work. All his ideas display these qualities. The mosaic he paints combine world history with events of his own life and the life of his ancestors. At the minimum, the story is a fantastic and a rich vision of reality; and I believe that a visionary, even if he is proven wrong in the long run, still transcends the mundane. It has a liberating influence not only on the visionary, but also on all those who witness it. How dull the world is without visions and visionaries.

A dream, if not the vision, of the Hubble telescope was already in the eyes of the first man who climbed down from the trees and stood on the edge of the savanna to look upward towards the stars. I believe there were already questions and longings in his primitive and curious mind. Already, there was hunger to know. Who am I? Where am I? Where did I come from? What is it all about Alfie?

I do not know if there are definitive answers to these questions. But ask we must. This is what makes us human. The questioning mind is our hard-earned heritage. It is the curse as well as the gift of being conscious, the price and the prize of self-awareness. We have not fully answered any of these questions yet. Perhaps we never shall; or there may be too many answers to be meaningful for this short life. There may be as many answers as there are names of God in the Cabala. Or perhaps we are orphans after all, a meaningless happening in a meaningless universe. Even if that is so, one thing is clear: just by asking the questions, we have greatly benefited as a species. Because of his existential uncertainties combined with a sort of stubborn fortitude to overcome them, Man has conquered nature and the night, though the uncertainties still remain.

Let us take the example of the night and the dark it brings. I do not believe that our ancestors ever cowered silently during the dark nights, hiding from the saber-toothed tiger or whatever else. We, humans are too noisy and quarrelsome a lot for that sort of behavior. We are a species of chance takers. If we are orphans of an indifferent universe, we are orphans with the courage to dream. Still the night was a fearful time and we have always wished to dispel the dark. However, when Ben Franklin flew his kite in the storm, light bulbs were not on his mind; nor were they on Michael Faraday's when he was trying to understand the nature of electromagnetism. These men just wanted to know. A byproduct, just one of the many, of the electromagnetic technology that came from this curiosity is that we have banished the night from human concerns as surely as we have banished the tiger. In both cases, I am afraid, with much too much vehemence and success. The tiger is on the way to extinction and the astronomers are complaining that the ambient light is blinding their telescopes and obscuring the skies. These are the fruits of a mere detour on the road to want to know who we are. Let the search continue. It is always better to know.

The impulse to search for the meaning of the universe seems to be intimately entwined with the search for the meaning of personal existence; and I suspect the answer, if any, may be one and the same or, at least connected. If Jack Sarfatti puts himself center stage and co-mingles the significance of his own existence with that of the universe, it is not necessarily a reason to reject his view. It is an old and respectable tradition.[140] The venerable Upanishads unabashedly equate

---

[140] Walt Whitman "Song of Myself" http://www.americanpoems.com/poets/waltwhitman/

the self with Brahma- the seed and the flowering of all existence. They make no distinction between the individual consciousness and the collective consciousness other than that of personal attitude.

Though the questions remain the same, much has changed since the times of the Upanishads. The sages of the Upanishads relied on poetry and parable to communicate their insights and carried no burden of proof. We moderns, however, insist on the rigorous scientific method communicated by the unambiguously neutral language of mathematics and supported and confirmed by repeatable experiments. This has paid off—at least as far as the understanding of material world is concerned.

I will not say that science has become our religion. It has not. In all honesty, neither has it tried to. Religion is expected to speculate on the hereafter. Indeed all religions offer explanations of death and create scenarios of what happens after the Big Event. It can be reincarnation; it can be resurrection; it can even be an everlasting limbo of the Ancient Egyptians, a sort of a capitalist's heaven: comfortable if you are Pharaoh, ignoble if you are fellahin. Science, other than a relentless and essentially unsuccessful effort to prolong life, has assiduously stayed out of this debate. "Death is not an event in this life", said Wittgenstein, a true genius child of the scientific mind-set. Scientists, as much as they may suffer the existential anxiety shared by all mortal men, have, as scientists, restricted themselves to study only this life and have kept clear of the hereafter. They have also stayed detached from any search for possible meaning of our fleeting existence.

There is a fly in the Bell Jar of this ointment of detachment, however, and that is the phenomenon of consciousness. All questions about the material world eventually hark back to the self. Who is asking? Physics has answered many questions about our physical world, so many in fact that we can blow up our planet many times over, and may yet do it from sheer cussedness. This is flirting with nothingness and as Nietzsche said, "When you look into the abyss, the abyss looks back at you".

It is not just the fear of self-annihilation alone, however, that is leading the scientists to study consciousness. The very success of objective science is forcing it to face the subjective. For example, in the epistemology of quantum mechanics, one of science's most successful systems, the observer plays a central role. It is not so surprising then that an increasing number of physicists want to know what makes us tick. The need to understand the watcher himself, the one who is asking the questions is becoming essential even to scientific inquiry. Will the scientific method be as successful in solving the mystery of consciousness as

---

http://www.americanpoems.com/poets/waltwhitman/songofmyself.shtml
Norman Mailer, "Advertisements for Myself" http://www.bartleby.com/65/ma/Mailer-N.html

it had been in explaining the material world? The task, it must be remembered, is of a different magnitude when the subject is consciousness.

Jack Sarfatti is one of these researchers. He has devoted more than 30 years to the study of consciousness and offers an eloquent model to his fellow physicists. Not surprisingly, it is controversial. But boldness, even a little recklessness is called for. This territory was not only uncharted but forbidden to the scientist. It was a place where formidable paradoxes loomed like dragons. The best of philosophers have lost themselves in the quagmire of arguments about consciousness. Religion is the only foolhardy force that wanders freely into this sticky thicket and has, as a result, conjured as many wondrous scenarios as the human mind can bear. Science, the most powerful fruit of the human imagination has been simply mute on the subject, and this includes quantum mechanics. "From the point of view of quantum mechanics, the faculty of self awareness is completely unexplained," lamented the late Prof. Eugene Wigner. (Eugene Paul Wigner was the first among the dons of physics to speak publicly and publish on the subject of consciousness and the role that physics may play in it. I came across this minutia when during a conversation, Jack Sarfatti, with an uncharacteristic modesty, refused to take the credit of being the first to do so. ("It was Eugene Wigner!" He was already talking and writing about it when I first became interested in the phenomenon. "You should read him." He told me with some passion and uncharacteristic modesty.)

I am not a physicist. Therefore, I cannot defend the technical aspects of his theory, but I can cheer him on for its eloquence. His cosmology, as he translates it into layman's language, has a great appeal to me. His concept of 'back-action' feels as solid as Newton's Third Law and still has a reach towards the unknown. The reoccurring symmetry it implies, a sort of ethical invariant that permeates all levels of reality is pretty as a poem. Sarfatti, of course, offers more than mere poetry. He offers a coherent and compelling model of consciousness. When he speaks of it in plain English, as he must to non-physicists like me- his model has an inner consistency and eloquence. Plain English spoken with clarity is nothing to scoff at. Language is, after all, the mother of mathematics. There is another reason I find his cosmology congenial. Sarfatti, despite his cantankerous persona, is a reconciler. He sees similarities in disparities and seeks patterns in randomness.

I belong to the Borgesian 'any description of Reality only adds to its complexity' school. I leave it to his fellow physicists to prove him right or wrong. It is enough for me that his theory is dynamic in its premise and inclusive in its vision. I admire his attempt to put his intellectual arms around the whole universe inflated into what he has coined as "Super Cosmos" and not just an itty-bitty portion of it. And I hope there really is quiet in the heart of chaos and stillness in the swirling motion.

Contact by Saul-Paul Sirag

The Yeshiva Boy asks Rabbi Sirag Four Questions[141]

"Rabbi, why does anything exist? Does the universe create itself? Does the universe need God? How is Jack's theory different from all other theories?"[142]

The Godfather made an offer Carl Sagan refused

It is well known that the movie version of *Contact* was based on the novel *Contact*, by Carl Sagan (published in 1985). Not so well known is the fact that the original version of *Contact* was supposed to be a mini-TV series produced and directed by Francis Ford Coppola.[143] And thereby hangs a tale.

---

[141] "The Child" Andrew of "The Woman"

[142] "Why do we eat unleavened bread? Why is this Knight different from all the others?"

[143] Francis is today executive producer of "First Wave" on the Sci-Fi Channel about ET invasion of Earth. The schism in the UFO community today is that of the "Alien Huggers" vs. General Douglas MacArthur's position. Also see "Hollywood Vs the

The protagonist of the movie and the book is the astronomer Ellie Arroway. In the book she is described as driving a red sports car with the bumper sticker: "Black holes are out of sight." Now the first time that Jack Sarfatti and I saw that bumper sticker was on Ellie Coppola's red sports car. So for us Ellie Arroway was a character who was a mixture of Ellie Coppola and Jill Tarter (an astronomer who has devoted her career to the SETI [Search for Extra Terrestrial Intelligence] project). Jack met Ellie Coppola in 1975 at the April-C est training by Werner Erhard. Ellie was carrying a copy of the book *Space-Time, and Beyond*, (by Bob Toben, Jack Sarfatti, and Fred Alan Wolf)[144] which had just been published by Dutton.[145] So Jack got to know Ellie and Francis, and began introducing them to his friends and contacts in San Francisco (and vice versa). In this way the Coppolas met Uri Geller (whose Spectra contact story intertwines this story) and Jacques Vallee, who became a consultant to Francis' friend Steven Spielberg on the movie *Close Encounters of the Third Kind*; and in this movie Vallee was the prototype for the character played by Francois Truffaut.[146] This meeting between Coppola and Vallee was in 1976, the year that I was hired by the Coppolas to tutor their son Gio in mathematics and physics. As basic reading material I used the book, *Intelligent Life in the Universe*, by Carl Sagan and I.S. Shlovskii. This was Carl's first book (published in 1966) and he was sympathetic to the idea of possible ET contact in the period of the early civilizations in Mesopotamia (3500 BC). During that year (1976) I kept a notebook of ideas, called *Time Traveler's Notebook*, and later got a series of (ever more insistent) requests from a Russian in Novosibirsk for a novel on time travel I was supposed to have written. (I haven't written it yet, but maybe I will!)

In 1985, when Sagan was writing *Contact*, he wanted to have the most accurate description of the idea of passage through black holes to other places and times of the universe. So he contacted his friend Kip Thorne, the Caltech physicist (a former graduate student of John Wheeler, and co-author with Wheeler and Charles Misner of the seminal textbook, *Gravitation*). Kip told Carl that black holes were not the way to go but that wormholes might be feasible,

---

Aliens" by Bruce Rux and "UFOS and the National Security State" by Richard Dolan for more details. "The Star Gate Conspiracy" is also relevant but it has a bizarre political spin to it. Sarfatti comment

[144] The literary agent for this book was the notorious Ira Einhorn now in a Pennsylvania prison for the 1977 murder of his girlfriend, Holly Maddux. There are still lingering doubts about Einhorn's guilt.

[145] Jack took his name off later editions because of the Ira Einhorn scandal, but still continued to collect a third of the royalties.

[146] Who was working with Jean Reisser Nadal. Both Nadal and Truffaut died prematurely which Jack considers suspicious because of what they were working on at the time that Jack was privy to.

provided they could be kept open with exotic matter.[147] He put a couple of graduate students (M. Morris and U. Yurtsever) on the project, with the result that new ideas about time machines were published, creating a flurry of interest in the whole issue of time-machines. (See *Black Holes & Time Warps*, by Kip Thorne, 1994, where Kip tells this story).

Jacob Atabet at the End of Ordinary History

Less well known than the Kip & Carl story, is the fact that in 1977 when Michael Murphy (the co-founder of Esalen Institute) was writing his novel *Jacob Atabet* I was tutoring Mike in physics and told him about Einstein-Rosen bridges (another name for non-traversable wormholes) with event horizons. [148]

[Note: I had learned about Einstein-Rosen bridges from the book *Gravitation* (1973) by Charles Misner, Kip Thorne, and John Archibald Wheeler. The picture of such a bridge is on page 837 and is taken from the 1962 paper by Wheeler and his graduate student Robert (Bob) Fuller, "Causality and multiply-connected space-time," *Phys. Rev. 128*, 919-929. By 1975 Bob Fuller had become a close advisor to Werner Erhard. He organized physics conferences in San Francisco,[149] was a director of the est Foundation, and was later put in charge of Erhard's *Hunger Project.* Fuller resigned from that project, when Erhard spent a large amount of money supporting himself as a race-car driver. Jack and Fred and I met Bob Fuller, at various meetings, parties and dinners, when we became, for a short time, science advisors to Erhard, teaching physics to his *est* trainers.]

---

[147] This antigravitating exotic matter is a positive cosmological vacuum local $\Lambda(\mathbf{x})$ field in Jack's new theory in 2001, with dark matter as a gravitating negative $\Lambda(\mathbf{x})$ field.

[148] "One way passages surrounding a spacetime singularity. If you go in along a free float geodesic, you cannot get out provided you are real and not quantum virtual, and move at or less than the speed of light. If, for smaller black holes of about ten solar masses or less, you attempt to hover on a nongeodesic you die from stretch-squeeze tidal forces (e.g. 15 g at 20,000 km above event horizon of 10 solar mass non-rotating black hole) as you get closer to the event horizon whilst still outside it without reaching the singularity where the same fate awaits you. None of this happens in a star gate, which is a traversable wormhole *without* an event horizon and without an interior singularity." - comment by Jack Sarfatti

[149] With Wheeler, Feynman, Hawking, Susskind, Gell-Mann, Finkelstein, Coleman and many others attending.

This Einstein-Rosen bridge idea[150] got into the Michael Murphy's *Jacob Atabet* novel (p. 148), which was set in an apartment in North Beach, formerly occupied by Mike, who had handed it on to Jack Sarfatti. To some extent, Jacob Aatabet is an artist modeled on Jack. So, in a strange way, as Ellie Coppola is to Ellie Arroway as Jack Sarfatti is to Jacob Atabet; while I am to Mike Murphy as Kip Thorne was to Carl Sagan.

Uri Geller meets Spectra

My interest in time-travel goes way back, but was especially piqued by my contact with Uri Geller and Andrija Puharich. in 1973, when I was writing an investigative story on Geller for *Esquire* (too controversial for *Esquire*, as it turned out). Geller and Puharich said that they had been contacted in various ways, but especially (in 1971) by voices on tapes (which disappeared) claiming to be Spectra a computer aboard a spaceship from the future. http://stardrive.org/cartoon/spectra.html When Puharich's book *Uri* came out in 1974, Puharich gave a copy to Jack, who immediately passed it on to his mother, Millie.

---

[150] The original Einstein-Rosen wormhole cannot be used as a stargate. It pinches off crushing the traveler inside it. It was not until Kip Thorne's solution that the idea became plausible. To repeat, because it is new and important if true, Jack's new local cosmological field idea has an exotic phase of the quantum vacuum $\Lambda > 0$ that permits Kip's solution and also Alcubierre's warp drive to be achieved as a matter of principle.

"A Thoroughly Modern Millie"

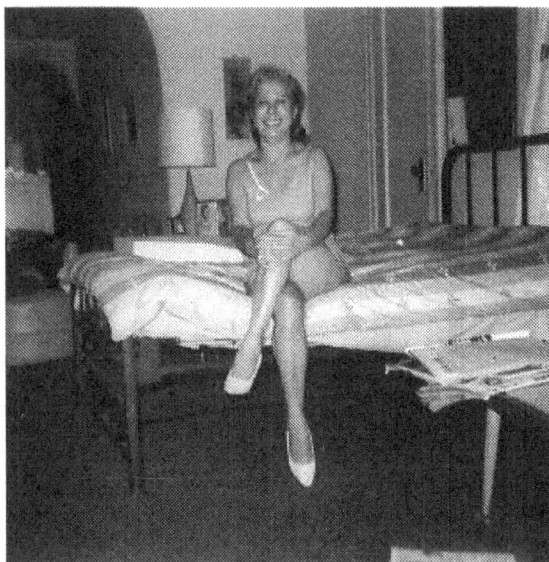

Jack's mother in 1953 where the phone call (s) came.

Jack at time of the phone call (s) in 1953 in Flatbush, Brooklyn.

When Millie came to the part about Spectra, she recalled three weeks of phone calls that Jack had received in 1953 from a metallic voice claiming to be a computer claiming to be aboard a spaceship. Jack remembered only one phone call—so, in effect, his memory had been partially erased.

Ezekiel's Vision

By April of 1974, I had heard about Jack and had seen one of his papers, which had a reference to a time machine effect. I was then writing the "libretto" for a very avant-garde opera called *Ezekiel's Vision* to be performed at the University Art Museum on June 7. I actually used a few lines from Jack's paper in the opera; and Tom Buckner, baritone, actually sang these words:

"We are creating the Cabala right now. General relativity provides 'time machines' in the form of 'closed time-like curves.' It would be possible for a super-conscious culture to go back in historical time and create its own history on one of the space-time pages in the great book of the cosmos." [151]

The idea of the opera was that what Ezekiel (cf. *Ezekiel*, chapters 1 & 10) saw in his famous Cherubim - Merkava vision was a Cabalist in a space-time vehicle going back to Babylon to see Ezekiel's vision. Rather than merely seeing the vision, he is part of the vision; and his efforts to see the vision help to create the vision. Notice that this is a self-consistent time-machine effect. That is, although the time-traveling Cabalist visits the past to share in Ezekiel's vision, he does not change it because he is part of Ezekiel's vision. This is the self-consistent type of time travel to the past described by the Russian physicist, Igor D. Novikov, *The River of Time* (1998). [152]

[Jack has just met Novikov at the Festschrift for Kip Thorne at Caltech, in June 2000; and at this Kip Fest, Novikov described his idea of self-consistent time travel to the past.[153]]

The Minds of Robots

In May of 1974, I attended an Esalen sponsored conference in San Francisco, in which the physicist Nick Herbert (whom I first met at the Institute for the

---

[151] Jack wrote this several years before Kip Thorne wrote of an "advanced technological civilization" amplifying tiny wormholes into star gates and years before the many worlds or "pages" of Stephen Hawking's "O Brane New World" in hyperspace.

[152] Jagdish Mann uses this idea in his romantic fiction "Nefertiti's Eye" in this anthology.

[153] This is also partly what Jack means by "Destiny Matrix".

Study of Consciousness in October 1973) was giving a talk on the physics of consciousness. He cited two examples of the type of theory he was interested in developing: Evan Harris Walker's theory in which the hidden variables of quantum theory are identified with consciousness ("The nature of consciousness," *Mathematical Biosciences* 7: 131-178, 1970); and James T. Culbertson's space-time network theory[154] in which memory is not located in the brain but in space-time itself, the brain being merely the instrument accessing the memory (*The Minds of Robots*, 1963). During the question period, I mentioned that I had recently read a paper by Jack Sarfatti, which was a more cosmic model of consciousness involving the future affecting the present. This created a big stir, and the biologist Brendan O'Regan (who was involved in the Geller research) spoke up and said that he knew Jack quite well, and that Jack had written a paper for the first issue of the journal *Psychoenergetic Systems*, of which Brendan was a staff member. I went to lunch with Nick, Brendan, and the physicist, Helmut Schmidt (who was doing research on retro-psychokinesis). Brendan told me that I would soon be meeting Jack—whether I could take him emotionally or not. Synchronistically, as we were walking along Powell Street to a restaurant, we passed a newspaper in a news rack, with the headline: "Helmut Schmidt: now German Chancellor." I turned to Helmut and said; "I guess you are now the Chancellor of West Germany!" But, of course, Helmut was more interested in science than politics, so he laughed it off. (This is but one example of the strange doubling of names in this story.)

---

[154] Roger Penrose's "spin networks in the pre-geometry" basis for loop quantum gravity - Sarfatti comment.

Nick Herbert, Saul-Paul Sirag in Vulcan Mind Meld ☺ induced by Jack Sarfatti ~ 1994

On September 14, of 1974 (his 35th birthday, two weeks after my 35th birthday), Jack showed up at the Institute for the Study of Consciousness in Berkeley, where I had been the live-in research assistant to Arthur Young since August of 1973. Arthur (the inventor of the Bell helicopter) and founder of the institute had been studying strange phenomena for many years. He had worked with Puharich in the 1950s, and was the mentor to many people, including Robert Temple, who went on to write *The Sirius Mystery* (published in 1976). Arthur[155] was very supportive of the research work with Geller, and had told me about Geller in 1972, when I attended a seminar he taught in the psychology department at UCB on his first trip to Berkeley. The graduate student, Ken Pelletier, who convinced Arthur to set up the Institute for the Study of Consciousness in Berkeley in 1973, sponsored this seminar.

The reason I knew about this seminar was that Jeffrey Mishlove, a UC graduate student working on an M.S. in criminology (and later received a Ph.D. in parapsychology from UCB), mentioned it to me as possible material for a

---

[155] Arthur Young was a key patron of Ira Einhorn who would visit the Institute in Berkeley several times.

weekly column I had been writing since August 1971: *The New Alchemy*, on the far-out developments in science. It was because of my column that I had met Jeffery on his birthday in December 1971, and we had become good friends after running into each other again at a UFO lecture by Stanton Friedman, on the Berkeley campus in early 1972. I was always on the lookout for new material for my *New Alchemy* column; and shortly after meeting Arthur Young, I came across the book *Breakthrough to Creativity*, by Shafica Karagula, a psychiatrist who studied people with positive abnormalities—such as the ability to learn advanced physics by attending a "dream school." Of course, I wanted to attend such a school! So I suggested to Jeffrey (who was also very fascinated by the book) that we try to get Dr. Karagula to give a lecture at UC. Jeffrey quickly extended that idea to organizing a conference of speakers to be sponsored by the Student Union on the UCB campus.

This conference occurred in March 1973; and among these speakers was Andrija Puharich, whose entire talk was on the research with Uri Geller. This created a great hubbub in Berkeley, and various experiments were planned for the expected arrival of Geller. He came in April and did his public demonstrations at Zellerbach Auditorium—telepathy PK ring bending and breaking, watch-fixing and so on. Before the show, he did a blindfold drive in a van full of people (including Alan Vaughan), through the winding roads of the Berkeley hills. Then there were smaller demonstrations, for instance in the physics department. This was not so spectacular. However, Uri did bend the wedding ring of one of the professors, who was impressed at first, but was talked into saying that he must have been hypnotized into a distracted state. He left on a trip to Alaska the next day. There were several other tests at various locations in the Bay Area. Many accused Geller of fraud, i.e., the tricks of an accomplished magician[156]. Many others were impressed by at least some of the phenomena. For example, a crystalline structure could be cracked by Geller and then subjected to electron microscopy, and compared with a normally cracked crystal[157] of the same kind.

The most documented research with Geller was at the huge research complex, SRI in Menlo Park, near the campus of Stanford University in Palo Alto. Hal Puthoff and Russell Targ, beginning in 1972 and continued in 1973 carried this out. See, for example, *Mind Reach* (1977) by Puthoff and Targ.

---

[156] Martin Gardner's "Magic and Paraphysics" in "Science, Good, Bad and Bogus" is a long article about both Jack's early immature theory and Uri.

[157] Hagen Kleinert of the Free University of Berlin shows that Einstein's explanation of gravity as local curvature corresponds to stringy cracks in the 4-dimensional "world crystal" that coherently freezes out of the quantum gravity foam in Jack's new theory.

Ghost Busters

Jack, Saul-Paul and Fred Francis Ford Coppola's CITY, 1975

It was in the summer of 1973 that Jack got involved with the Geller phenomenon. It happened like this. Jack was teaching physics at San Diego State University. He had just broken up with his wife, and was living with another SDSU physicist, Fred Alan Wolf, who had also recently split up with his wife.

Fred's old high school buddy, Bob Toben, arrived from Chicago out of the blue, with a book contract. Bob was fascinated with the Geller phenomenon, and wanted Fred (and Jack) to explain it. Bob tape-recorded their wild speculations; he transcribed this material; provided cartoon-illustrations; then had Jack rework the explanations of the illustrations and provide an extensive annotated bibliography. Dutton published this in 1975 as the book *Space-Time and Beyond.* (This was the book, as mentioned previously, that Ellie Coppola was carrying when Jack first met her in April 1975 at Werner Erhard's "Celebrity Training".) But, back in 1973, shortly after Jack first heard about Geller from Bob Toben, he was visiting his girl friend's mother in Carmel Valley. He was about to go off to Trieste, Italy to work with Abdus Salam (a Nobel laureate by December of 1979) at the International Center for Theoretical Physics . He had only a few days to spend in Northern California. The *Sunday Chronicle* came out and in the magazine section, was a story (by Gail Sullivan) about the new parapsychology work in the Bay Area. Especially prominent was the account of

Puthoff and Targ's research[158] with Geller at SRI. Jack immediately called SRI, and talked to Brendan O'Regan, who had already heard about him, and invited him up for a visit—which lasted 17 hours. Then Brendan drove Jack to Dean Brown's house in Portola Valley, where a small group of scientists (including Hal Puthoff and Russell Targ) had gathered to discuss Geller and especially the ET-contact-with-Spectra part of the Puharich-Geller story. Jacques Vallee had suggested the meeting but was not present. Curiously, Vallee, had himself spent a year working on RCA's Spectra 70 computer, before RCA completely got out of the computer business in the fall of 1971. (See *The Network Revolution: Confessions of a Computer Scientist*,[159] by Jacques Vallee, 1982.)

[Incidentally, a friend of mine, a graduate student in computer science, rescued a bunch of Spectra 70 computer manuals from a dumpster on the Berkeley campus, after UC got rid of its Spectra computer. I showed one of these manuals to Geller, and he was quite amazed! -"Andrija, look at this, look at this!" is what he said.]

Back to Jack in 1973: at his first meeting with the group of scientists who were actively researching the Geller phenomenon, when the Puharich-Geller story of the voice (calling itself Spectra) on the tape recorder came up, Jack spoke up and briefly told the story of his 1953 phone message of a cold, metallic voice claiming to be aboard a spaceship. The voice said that he had been selected as one of 400 bright, receptive minds and that he would be meeting some of the others in 20 years. It is important to note that Jack, in 1973 (and to this day) remembered only one phone call. But as mentioned previously, when his mother read Puharich's book *Uri* in 1974, she remembered three weeks of phone calls that Jack was involved with in 1953. She was struck by the similarity between the description of the Spectra voice described in *Uri*, and the voice she heard on the phone, when she ended the series of calls by grabbing the phone out of Jack's hand and yelling into the phone, "You leave my boy alone!"

[Note: the exact wording of Jack's story (in which he describes only *one* phone call[160]) is on a tape recording made by Frieda Morris at this 1973 meeting. She later gave the tape to me. It is interesting that this crucial piece of Jack's story is preserved in this manner. This was not a disappearing tape!]

---

[158] Funded by the Central Intelligence Agency. This was not public information back in 1973.

[159] This Vallee book had a description of Werner Erhard's est in the beginnings of the Silicon Valley computer industry. Ira Einhorn claimed he was working with Vallee at that time when he met Jack and Kim in Bolinas after his indictment for the murder of Holly Maddux. Ira had just driven north from Esalen in Big Sur.

[160] Jack's mother, Millie, would not read Andrija's URI for another year. It had not even been written then.

Jack went off to Trieste to work with Abdus Salam right after this seminal meeting in Portola Valley. The SRI scientists continued to work with Geller; and Puthoff and Targ published a paper on this research in the October 18, 1974 issue of *Nature* ("Information Transfer under Conditions of Sensory Shielding").

Sharon and Jack in Venice, 1973

It should be noted that even though I had spent much time with Geller and Puharich in the spring of 1973, I had learned only about the disappearing messages (claiming to be ET) on tape recorders. There was also the red light

101

version of ET, saying, "Now you can see what we look like." And there were pictures of disk shaped UFOs that Uri had taken from airplane windows. I had yet to hear about Spectra and the hawk-like ET presence. [161]

Horus, the Egyptian Sky God through the Star Gate?

Then in the fall of 1973, I began to hear stories from Ray Stanford in Austin, Texas, about an entity he called Spectra teleporting Ray's car (on two occasions) while he was driving to the airport to pick up Geller. With these stories in the back of my mind, I was really shocked when the January 1974, issue of *Analog* hit the newsstands in mid December 1973. There on the cover, was the picture of a man in a white uniform, with a hawk-like helmet. His nametag read "Stanford;" and the title of the story was "The Horus Errand." Note that Horus is the ancient Egyptian hawk-headed god.

The story had to do with reincarnation in a futuristic Los Angeles, where agents, such as the hero Stanford, could guide the soul of a dying person into the body a pre-selected newborn baby. Now Ray Stanford (who bears a striking resemblance to the Stanford on the Analog cover) does believe in reincarnation, and so (as it turned out) did the illustrator of the story, Kelly Freas. He stated in a letter to Alan Vaughan (articles editor of *Psychic Magazine* at that time), that he had never met Ray Stanford, but ten years earlier had had a psychic reading done by Stanford (by airmail); for while Stanford lives in Austin, Kelly Freas

---

[161] See also Chapter 11 of "Remote Viewers: The Secret History of America's Psychic Spies" by Jim Schnabel. Colonel John Alexander told Jack that this book is reliable. The story in Chapter 11 is similar to the one here. It involves CIA's Kit Green, Harold Puthoff, Russell Targ and physicists from Lawrence Livermore Laboratory (LLL) checking out Uri Geller. Some of them were part of Elizabeth Rauscher's "Fundamental Physics Group" at Lawrence Berkeley Laboratory (LB). Green, at that time, had the same job Ron Pandolfi later took. Harold Chipman's role in this is mysterious. Jack Sarfatti wrote: "Indeed, former high-level CIA officer Kit Green, MD and former NSA employee Hal Puthoff, both confirmed to me (by e-mail in 2002) the essential authenticity of this story that is clearly similar to my 1953 contact experience. Green's and Puthoff's testimony corroborating the very uncanny bizarre "metallic voice tape" (and the "one armed man" banging on the door) story, told in Chapter 11 "You can't go home again" of James Schnabel's "Remote Viewers: The Secret History of America's Psychic Spies", is important evidence of post-quantum objective weirdness with retroactive advanced signal-nonlocality in obviously paranormal human events of immense historical important (e.g. Churchill's and Mussolini's childhood "remote-viewing" precognitions, the alleged precognition of 911 by Usama Bin Laden's terrorist followers etc.) and the plausibility of my own 1953 testimony. Colonel John Alexander (USA, ret) author of "Future War" and former head of Los Alamos National Laboratory's "Nonlethal Weapons" program, also corroborated Schnabel's book in general to me about two years earlier. To stick our heads in the sand in denial is not wise."

lived in Virginia Beach, Virginia (because of his interest in the Edgar Casey research group located there). Kelly said that this psychic reading suggested that he had lived previously in ancient Egypt, where he was an illustrator of some kind. For this reason, he exaggerates any Egyptian symbolism mentioned in the stories he illustrates. In the case of the Stanford *Analog* cover, he painted the face from imagination (although he usually painted faces from photographs). This would seem to be an example of remote viewing, which was at that time being investigated by Puthoff and Targ at SRI.

[Incidentally, the non-fiction article in that January 1974 issue of *Analog* was, "Space Probe from Epsilon Bootes?" by Duncan Lunan; so "ET" was "in the air" very strongly at that time.]

The main reason that this hawk-headed Stanford was especially meaningful to me was that I had attempted to contact Uri's ET presence in June 1973, while I was working on the *Esquire* story. Andrija and Uri had told me about the ET voice on disappearing tapes. They had also mentioned a red laser-light version of Geller's ET presence. They had not mentioned the Horus hawk version of the ET presence (or the name Spectra). My method of trying to contact their ET presence was to spend an afternoon with Geller, on my own turf (in a friend's loft in Manhattan) while I was in the psychedelic state induced by LSD. Puharich was sympathetic to this approach, since he had done extensive work on psychedelics in the 1950s, including work for the military (see *The Sacred Mushroom* (1959) and *Beyond Telepathy* (1962) by Andrija Puharich). During this trip, at a moment I felt was right, I asked if I could make contact with his ET presence. He told me to look into his eyes and tell him what I saw. I thought to myself, that I would probably only see some red laser-type light in his eyes, since I had already been told Uri's mysterious red laser story. But I was very surprised to see not only his eyes, but also his entire head take on what I took to be an eagle shape complete with feathers going down to his shoulders. I jumped back a step and said, "Uri, you look just like an eagle." He was very excited about this, but wouldn't reveal anything further about the ET presence. When, Puharich's book *Uri*, with its extensive and detailed Horus hawk stories, came out later in 1974, I understood why Uri had been so excited by my seeing him as an eagle while trying contact the ET presence. The "Horus Errand" *Analog* cover story of January 1974 had already sensitized me to parallels with the Puharich-Geller story. In fact, one of Kelly Freas' Egyptoid touches on the cover was the large pyramid shaped building in the background—with a red laser beam shooting out the top into the sky. The pyramid-shaped building was not in the story itself, but was part of Kelly's Egyptoid exaggerations—along with the hawk-headed helmet—described in the story as merely a white and gold helmet. But the red laser beam shooting off the pyramid building was not in the story either, and could not be considered Egyptoid. It was close to what I had expected to see in Uri's eyes when I saw his head suddenly become eagle-like. It is because of experiences like this that I take the ET part of the *Uri* saga more seriously than many other readers of that book.

Young Frankenstein

Jack Sarfatti & Fred Alan Wolf in front
Saul-Paul Sirag & Nick Herbert in back, 1975
Photo appeared in Francis Ford Coppola's CITY Magazine in 1975
In "Faster Than The Speeding Photon" by Rasa Gustaitus

http://www.tmbhs.com/tmbhs/movies/youngfrankenstein/youngfrankenstein.asp

The usual suspects 25 years later at Russell Targ's house.

Bell, Book and Candle

By January of 1975 Jack and I were setting up the Physics Consciousness Research Group to conduct seminars on the new developments in physics, which were relevant to the understanding of consciousness. The main thing we discussed was quantum theory, and the recent work on Bell's theorem. I had first heard about Bell's theorem from the physicist, Nick Herbert, when I met him at the Institute for the Study of Consciousness in October of 1973. Bell's theorem (1964, republished in *Speakable and Unspeakable in Quantum Mechanics*, J.S. Bell, 1987) was a simple mathematical proof that the standard predictions of quantum mechanics for a particular type of experiment (called an Einstein-Podolsky-Rosen experiment (EPR) -) were inconsistent with a description of this experiment that was both objective and local. Here "objective" means observer independent, and "local" means that no influences can go faster than light or backwards in time. The EPR experiment uses two particles that interact with each other, and then separate. According to quantum mechanics these two particles will continue to be correlated after their separation. Most people confronted with this statement, say, "Of course they will be correlated because they separated from the same place—it's as if I separated two sides of a coin and sent them off in opposite directions. They would continue to be correlated in the sense that if I found the head half in one location I would know immediately that the tail half was in the opposite location." In Bell's terminology, this would be an objective, local description. The point of Bell's theorem is that the quantum correlation for the two separated particles is much stronger than can be accounted for by such an objective, local description.

It is also important to realize that the EPR type of experiment has been performed in several different laboratories, first in Berkeley by John Clauser and a graduate student, Stuart Freedman (*Phys. Rev. Lett.* 28, 938-41, 1972), and later more definitively by Alain Aspect (with P. Grangier, G. Roger, & J. Dalibard) in Paris, (*Phys. Rev. Lett.*. 47, 460-466(1981); 49, 91-94 (1982); 49, 1804-1807 (1982)). The latest development of the nonlocality weirdness of quantum mechanics, is the work on "quantum teleportation" (Anton Zeilinger, *Scientific American*, April 2000.) The role of John Clauser, then a mere post-doc and now a physics professor at U.C. Berkeley, in this entire line of experimental development was seminal. His experiment was the small beginning, which led eventually to an avalanche of experiments.

Elizabeth Rauscher, then a graduate student in physics at Berkeley introduced Nick Herbert and me to John Clauser in early 1974. We saw the second version of his experiment running, but the equipment was primitive and the data acquisition was very slow. Yet Clauser was able to extract very significant consequences from this experiment—in fact he was redoing an experiment, which Richard Holt had done for his Ph.D. thesis at Harvard. Holt had results, which contradicted quantum mechanics; and this was terribly ironic, because Holt didn't want these results, whereas Clauser would have been quite happy with them. But when Clauser redid the Holt version of the EPR-type experiment, he got results consistent with quantum theory, and he made a good guess as to what went wrong with Holt's experiment. At the time Elizabeth, Nick and I saw this experiment running it was still touch-and-go whether the predictions of quantum theory would hold up or not. Nick, ever the joker, asked Clauser whether there were any "hidden variables" hiding around his laboratory—because it's the local hidden variables[162] that are ruled out by Bell's theorem.

Werner Erhard

Werner Erhard Long Beach 1977

http://qedcorp.com/book/psi/hitweapon.html

---

[162] "The hidden variables in Bohm's quantum realism are "nonlocal". That is, you can think of an electron approximately as a point particle with a continuous classical path. However, it feels a nonlocal quantum bit potential of active nonclassical information. The electron's path, though continuous, need not be differentiable, so there need not be a conflict with Heisenberg uncertainty even at this sub-quantum level." - Jack Sarfatti.

Many months later in early 1975, shortly after their est-training, I introduced Jack Sarfatti and Fred Alan Wolf to Clauser. On the way up to the Berkeley campus, Jack and Fred were trying out their est-speak on me. The occasion was a talk on Bell's theorem by Costa de Beauregard, a French physicist[163]. He had been visiting Hal Puthoff and Russell Targ at SRI because he was interested in their remote-viewing research[164], and the research with Geller. Hal and Russell brought de Beauregard to Berkeley, and we all met in the Physics department. Clauser, always ready for a talk on Bell's theorem, came up from the Berkeley yacht harbor where he had been painting his sailboat. Moreover, because his experiment was completed (with results strongly in agreement with quantum theory), he was able to show us all through the insides of his experimental equipment.[165] We also saw that he had placed a sign over his equipment that said: "We have met the hidden variables, and they is us.—Pogo."

The French Connection

De Beauregard's talk had been about his view of the EPR-Bell's theorem situation, which he interpreted as an illustration of the future affecting the present. This fits Jack Sarfatti's idea of course.[166] De Beauregard went back to Paris and encouraged Alain Aspect to do the definitive version of the Clauser's experiment (with Clauser as a consultant) which required much more expensive equipment, and some clever devices. Thus Aspect's experiment took until 1981 to complete, and was published in 1982. Later on Aspect came to SF and stayed with Jack at his apartment in North Beach.[167]

The impromptu talk by De Beauregard was attended by a number of keenly interested physicists: Henry Stapp, a theoretician at Lawrence Berkeley Laboratory (LBL), and an expert on Bell's theorem; Geoffrey Chew, who was head of the theory group at LBL; Elizabeth Rauscher, a graduate student at LBL,

---

[163] De Beauregard was from the same Institut Henri Poincare in Paris as Vigier, but with opposite politics and opposite philosophy about the paranormal. Vigier was a former Communist who pooh poohs paranormal. De Beauregard a right wing aristocrat who believed in the paranormal.

[164] Role of French Military Intelligence here? Sarfatti comment.

[165] The detector efficiency loophole has recently been plugged for massive particles, reported in Nature 2001 by Philippe Grangier.

[166] Also it fits John Cramer's "transactional interpretation" of the quantum idea from Wheeler and Feynman. Sarfatti comment.

[167] Where he met Lawry Chickering. Michael Berry also visited Jack at the Caffe Trieste. Creon Levit was at that meeting between Berry and Sarfatti.

George Weissman, a graduate student of Geoffrey Chew, and Mike Nitschke[168] an experimental physicist involved in creating trans-uranium elements. This de Beauregard meeting was a major spur to the weekly seminar led by Elizabeth Rauscher at LBL, which began in May 1975 and ran on for a couple of years. This seminar was called the Fundamental Physics Group and was focused on the understanding of Bell's theorem and its implications. It was the genesis of several books: Gary Zukav's *The Dancing Wu Li Masters*, Nick Herbert's *Quantum Reality*, Fred Alan Wolf's *Taking the Quantum Leap*. Fritjof Capra, had already written (but not yet published) his book, *The Tao of Physics*. Fritjof was usually present at the seminar; and in fact the very first meeting was a presentation of the ideas of this book.

The Electric Kool Aid Acid Test

The Fundamental Physics Group intermeshed with The Physics Consciousness Research Group (founded by Jack Sarfatti), and the Consciousness Theory Group (founded by me in 1977), which began at the Institute for the Study of Consciousness in Berkeley, and from late 1977 to late 1979 met at Henry Dakin's Washington Research Institute in San Francisco. For twelve years, beginning in 1976 there were Physics & Consciousness workshops at Esalen Institute. The month-long meeting in January 1976 was led by Jack Sarfatti, and evolved into 5-day seminars led by Nick Herbert and me.

The January 1976 month-long workshop at Esalen (in Big Sur) led by Jack Sarfatti and me (as Sancho Panza) was a wild and woolly adventure. We invited various scientists interested in consciousness to participate for a few days at a time.

I will describe some events of one day during this workshop. Fred Alan Wolf and I took an LSD trip during an outdoor lecture on consciousness and gravity by Claudio Naranjo (an expert on psychedelics, who had recently returned from Chile where he had led a group of Americans to participate in Oscar Ichazo's Arica training.)[169] During the lecture, Fred wandered over to the edge of the cliff overlooking the ocean. I started to worry about how he was taking his first LSD trip. He looked over the edge and said with a chuckle, "Naughty, naughty! Don't go there!" I was relieved when he came back to Claudio's gravity lecture, rather than doing a gravity experiment! At lunchtime

---

[168] One of the LLL weapons physicists in the metallic UFO voice incident in Chapter 11 of Jim Schnabel's "Remote Viewers" with CIA's Kit Green? - Sarfatti comment.

[169] Indeed, it was Jan Brewer's tales of this to us at Esalen that prompted Jack to write the parody "Hitler's Last Weapon", http://qedcorp.com/book/psi/hitweapon.html. George Koopman investigated Ichazo and Arica.

Fred and I didn't feel like eating so we spent some time looking at my Tibetan tanka depicting Avelokitishvara, the Bodhisattva of compassion. Jack came back after lunch with two messages. One was a package for each of us (Jack, Fred, and me) containing a book from Alan Laiken, called *How to Get Control of Time in Your Life*. On the first page of the promotional pamphlet accompanying the book was written the question: "Are you making the best use of your time?" We all cracked up—Fred most of all, in a laugh I will never forget. Just as synchronistic was the other message: Jack had just received a phone call from Timothy Leary, who was still in prison. He wanted to talk to Jack and me because we had sent him a packet of writing in response to a questionnaire that was making the rounds in 1975: what is your view of Higher Intelligence?—And similar questions. Leary had been invited by Ken Kesey (author of *One Flew Over the Cuckoo's Nest*[170]) to edit a special edition of his small literary magazine *Spit in the Ocean*. Leary said that our material was the best thing he had received, and it was subsequently published in *Spit in the Ocean* #3, 1977. One of the pieces I had sent to Leary was a lightly fictionalized version of experiments run in 1975 by Nick Herbert on his "metaphase typewriter, a quantum-mechanical open-mike to the void." The experiments (done at the U.C. Medical Center in SF) were plagued by very intriguing synchronicities, such as the words "by jung," while Alan Vaughan was trying to influence (presumably by psychokinesis) the quantum source into typing out meaningful expressions associated with pre-selected target words. The "by jung" phrase did not fit the target word but it seemed like—well, a Jungian, synchronicity. "Where did that come from?" we all asked, and a passing lab technician pulled out a copy of *The Portable Jung* from her lab-coat pocket, and said calmly, "Maybe from here!"

Back at Esalen, on the afternoon of the Naranjo talk, Nick Herbert arrived with his friend Ralph Abraham, a math professor at U.C. Santa Cruz, who gave us a talk on and a demonstration of the newly emerging idea of chaos[171] and fractals[172]. (See James Gleick's book *Chaos).* Abraham also told us about Rene Thom's catastrophe theory, which has played a very important role in my physics work of recent years (See my paper, "Consciousness: a Hyperspace View," published as an appendix in Jeffrey Mishlove's, *Roots of Consciousness (2nd*.

---

[170] Jack Sarfatti saw the movie in Francis Ford Coppola's home with many of the people who made and acted in the movie.

[171] An instability of Newtonian classical mechanics in which a tiny shift in a starting point rapidly causes a large deviation from the continuous path that the particle would otherwise have taken. (Jack comment).

[172] Self-similar patterns of patterns like Ezekial's "wheels within wheels" and Russian Dolls. (Jack comment)

Edition), 1993.) Incidentally, Jack published a long appendix paper in the 1st edition of Jeffery's *Roots of Consciousness*, 1976. [173]

Shortly after Leary's call to us from prison, George Koopman called us from Southern California and wangled an invitation to Esalen. He seemed to know that Leary would soon be released from prison and that we had invited Leary to Esalen, and he wanted to meet Leary. He brought promises of funding to our newly formed group, the Physics-Consciousness Research Group, if he could participate from time to time. He told us that he had done his stint in the military by running the "nut desk" at the Defense Intelligence Agency. Question: was he still working for the "nut desk," and were we the new "nuts" on the scene he was checking out? He told us that the two highest priorities for investigation on the "nut desk" were UFOs and the USSR. George did fund us for a while. He participated in a couple of seminars, most notably the Esalen-Westerbeke Ranch seminar in the summer of 1976. After Leary was released from prison in August of 1976, Koopman befriended him and helped write one of Leary's books *Intelligence Agents* about "higher intelligence" which was part of Leary's SMI^{2}LE program: Space Migration, Intelligence Increase, and Life Extension. Koopman also wrote part of Leary's *Neuropolitics* (1977), and *The Intelligence Agents* (1978) Leary's motto for this program was: "The meek shall inherit the earth; the wise and strong move on." In consonance with this program Koopman formed the American Rocket Company. In 1986 he died in a single-car accident in the Mohave Desert on the way to a site where one of his rockets was to be demonstrated to prospective investors in his company. As a relative of the publisher of the New York Times, George Koopman was given a substantial NYT obituary. Koopman did not live to see the collapse of the Soviet Union or the much-increased interest in ETs and outer space, the two high priority concerns of the "nut desk." However, he was also interested in movies. He was a financial backer of *The Blues Brothers* and got to know John Belushi and Dan Akroyd. He claimed that he had convinced the special effects team to drop the police car (from an off-screen helicopter) onto the high-rise roof to make a "pin-point" landing—"a simple application of Newtonian mechanics" as he put it. He apparently used his experience with our Physics-Consciousness Research Group to inspire various aspects of the Dan Akroyd film *Ghost Busters,* which captures some of the wildness of those days in which various groups formed and folded and intersected in a frenzy of evolving thinking at the interfaces of science, art and religion.

---

[173] Which, like in *Dancing Wu Li Masters* and *Space-Time and Beyond*, disappeared in the later editions. Jack's articles, like Spectra's metallic voice tapes, have a tendency to disappear. ☺ (Jack's insert)

[174] *Physics Consciousness Research Group at the Westerbeke Ranch in Sonoma 1976*

These intersecting groups functioned somewhat like the contemporary rock music culture of Northern California. There was an infectious sense of joyously uncovering the nature of reality. This reality was seen, via Bell's theorem to be

---

[174] I'll name the people I recognize, starting from the back row, and reading from left to right (using "Blanck" for people I don't recognize: Blanck—Blanck—Dr. Frank Barr; Blanck—Blanck—Blanck—Emily—George Koopman (Allegedly of "Nut Desk" for DIA)—Blanck; Nick Herbert—Julia Kendell—Saul-Paul Sirag—Dorothy Kelly—Blanck; Betty Andreasson—Blanck—Anne Dale; Jack Sarfatti in the middle in white pants lying on the ground like a Pasha. This was at the Westerbeke Ranch, 2300 Grove Street, Sonoma CA. Koopman paid for this event that lasted for several days. Note on Emily: She is the woman to George Koopman's right. I don't remember her last name. But she came with George to Westerbeke. She ran an oriental rug store on Sacramento Street in San Francisco. George was a silent partner in the business, importing rugs from the Middle East (I don't remember which countries). But it might have been a cover for other activity—of course:-) I visited the rug store and Emily after Westerbeke. She was still there when I moved from Berkeley to SF in November 1977 to the flat (owned by Henry Dakin) on Washington Street with Barbara Honegger. The picture was taken with Emily's camera; and she gave me the print of the Westerbeke picture at that time. Note on Barbara Honegger: She was a White House staffer working for Martin Anderson in the early 1980s and is mentioned in Ron McRae's book *Mind Wars (St. Martin's, 1984)*. McRae, an ex Navy Intelligence officer working for Jack Anderson, stayed with Jack and Sam Sternberg in North Beach for a week while researching that book. Honegger wrote *October Surprise* (Tudor, 1989), after she left the Reagan White House. Barbara in 2002 is working for DOD in current Bush Administration and, as we go to press, has just blown the whistle on a Federal Judge she accuses of protecting al Qaeda operatives impeding the pre-911 FBI investigation according to http://www.gordonthomas.ie/122.html

"non-local" and by some of the more exuberant thinkers to entail a backwards-in-time effect. Jack especially championed this view. Could it be that the present wave of interest in this idea (such as at the Kip Thorne Fest in June 2000) was affecting Jack way back then?

John Archibald Wheeler, Jack Sarfatti, Cheryl Haley [175] ***at the Cal Tech Kip Fest, 2000***

---

[173] Close friend of Richard Feynman's at Esalen who aided him in his final days.

Cheryl Haley and Richard Feynman at Esalen, Big Sur, CA

Cheryl Haley with Jack Sarfatti at the Cal Tech Kip Fest 2000

Lynda Williams (The Physics Chanteuse), Jack and Cheryl at the Cal Tech Kip Fest

http://www.physics.sfsu.edu/grad/lwilliam/

A Biographical Note: Growing Up in a Japanese Prison Camp by Saul-Paul Sirag

I was born on Queen Wilhelmina's birthday on August 31,1939, the day before Hitler invaded Poland and started WWII, which was two weeks before Jack Sarfatti was born in Brooklyn. However, I was born in Dutch Borneo very far from either one of those events. My Dutch father and American mother were Baptist missionaries to the Dyaks, but I spent less than a year in Borneo. My earliest clear memories are in Java during the Japanese occupation. From December 1942 to September 1945, we were imprisoned in concentration camps. The men were imprisoned in camps quite separate from the women and children. My father died in a camp only 10 kilometers away from our camp near Ambarawa, which we learned only after the war. In the spirit of a time-machine to the past I will recount three experiences from that concentration camp period which strongly influence my present preoccupations: numbers, physics, and consciousness.

Every morning in front of our cell block, we would have to stand at attention and count off in Japanese—ichi, ni, san, shi, go, etc. up through sanjuichi, sanjuni, sanjusan, etc. (somewhere in the thirties). Sometimes, we would have to stand at attention for many hours, in order for our cells to be checked for contraband—for example, any books or pens or pencils. This every-morning counting in Japanese was how I learned to count; and I have continued to be fascinated by numbers. In the early morning, I used to count (mentally) before it was time to get up to see how high I could go. Sometimes I would get above a thousand, and that would "make my day."

From time to time I would ask my mother questions about the world outside the high cement walls of the camp Banju Biru (which had previously been a penitentiary). Especially important to me was to understand the overall structure of the Earth. Mother would tell me that the Earth was like a big ball, with our camp occupying a tiny little piece of it. I struggled and struggled with this concept, but found it impossible to clearly imagine such a thing. It seemed obvious that beyond the camp wall there must be a bigger wall holding in the island of Java, and beyond that a series of bigger and bigger walls, until finally there would have to be an immense wall holding in the sky. But the sky seemed to be too high to be held in by any such wall. So my wall model just didn't work. This to me is a parable of the necessity to go beyond preconceptions in our understanding of the universe.

Another experience was more directly related to the nature of mind. In the camp (of around 5000 prisoners) my mother, my brother and I were the only English speakers. Everyone else spoke Dutch (and some Malay—now called

Bahasa Indonesia). So Mother laid down a rule that we were not to speak English except privately in our one-man cell occupied by five people. English became a kind of secret, and sacred, language for Mother to tell Bible stories every night, which we had to repeat the next night before hearing a new story). This rule of no English except in private was hard to keep, so I would steel myself by playing a mental game that I called my "Yah-But" game. I would give myself a rule that I couldn't say anything. But since this was a game (in English) being played in my head, as soon as I assented to the rule by saying "Yah" (both English and Dutch for "yes"), I realized that I was breaking the rule, so I would say "But." And then I would see that I was breaking the rule again at a higher level. I would go through a whole series of "Yah-but-yah-but-yah-but-..." which could be punctuated as:

(((((yah) but) yah) but) yah) but...

And after each iteration of "yah-but" I would become more and more ecstatic, because I was breaking (and making) the rule at a higher and higher level. This is like the meta-levels implied by Goedel's incompleteness theorem, which I first learned about in 1963 from Mario Savio in another prison, the SF County Jail, after our participation (with hundreds of others) in the Sheraton-Palace civil-rights sit-in. But that is another story.

In any case, my "Yah-But" game was my secret method of "getting high". However by the time I was 7 years old I was living in an orphanage, called Christ's Home near Philadelphia. At that time I could remember dimly the emotional feel of the "Yah-But" game, but try as I might, I couldn't remember the simple rule for playing it. In 1965 I was in Berkeley experiencing my first LSD trip, and the whole experience of the "Yah-But" game came flooding back—with the childhood ecstasy generating rule: "I'm not allowed to say anything. Yah, but..." How could I not be fascinated with the nature of reality in all its guises, physical and mental?

The Men in Blue, a short surrealist playlet by Eugenia Macer-Story and Jack Sarfatti

Eugenia

Yesterday I was rushed, as there was "high strangeness" during a trip to NYC on Sunday. I wanted to check with you on one item. On the commuter bus leaving Kingston, NY at 10:30 a.m. there was an unusual gent who had been on the bus coming from Albany or Saugerties. When he got off the bus as I was entering, he looked me directly in the face and I thought: "That looks like Jack Sarfatti" He was wearing a cobalt blue beret. The man had at my first glance a trimmed gray beard like yours and was Jewish. Then after I had boarded the bus here-entered the bus and sat in the front seat a few rows in front of me across the aisle. I noticed that he was dressed entirely in bright cobalt blue. Blue long-sleeved, embroidered shirt; blue trousers; blue duck belt, matching blue shoes and a blue Walkman clipped to his belt with blue earphones, which matched his beret. The earphones seemed to be holding his beret on his head. Then, from behind, I noticed that he was wearing a curly peroxide blonde wig under the beret. His side hair in front of the ears and at the nape of the neck was dark and close-shaven. It was a very strange ensemble. Later in the day I had a "time dilation" experience, which I will refrain from discussing here. Also: my book of commuter tickets disappeared from my attaché case. I tell people it was pick pocketed. But I am not convinced the theft was conventional due to the time dilation.

Jack Sarfatti: That explains everything, especially the blonde wig under the blue beret. You see my dear Eugenia I wear Francis Ford Coppola's yellow beret to cover my blue blood. This is clearly Moriarty's cult "The Men in Blue"[176]

---

[176] Also "Brother Blue Resonant Human" of the "Sacerdotal Knights of National Security" who haunts cyberspace like "The Phantom of the Opera", perhaps more like Groucho Marx in "A Night At The Opera".

started by George Koopman, Dan Akroyd and John Belushi. They are servants of the "One Armed Man". We had a Beatnik Lady of The Saloon in North Beach named "Blue" for many years.

Jack Sarfatti: Can I put this in my book free of charge?

Eugenia Macer-Story: Certainly. You might add that I am an off-off Broadway poet/playwright and was on the way to view a colleague's play in a performance at the 72nd street Central Park outdoor stage.

Star Gate

Introduction to the new physics of macroscopic quantum geometrodynamics

Generic static spherically symmetric spacetime without horizon

This is a preliminary Thorne-Morris toy model[177] to illustrate the basic idea. It is not yet a star gate time machine.

The metric form in the usual Schwarzschild distorted spherical polar coordinates useful for a timelike nongeodesic LNIF[178] observer "fixed" relative to " $r \to 0$ " is

$$ds^2 = -e^{2\phi(r)}(cdt)^2 + \frac{dr^2}{\left(1 - \frac{b(r)}{r}\right)} + r^2 d\Omega^2$$

(1.1)

Where $d\Omega^2$ is usual flat space squared spherical polar angle element for latitude and longitude on the celestial sphere centered at $r \to 0$

$$d\Omega^2 \equiv d\theta^2 + \sin^2\theta d\varphi^2$$

(1.2)

The *anti-coarse grained* k resolution scale-dependent[179] macroscopic quantum corrected local Einstein geometrodynamic field equation is

---

[177] "Lorentzian Wormholes", Matt Visser, 2.3.12 (AIP Press). See Ch 11 for the next step.

[178] Local Non Inertial Observer

[179] The range of $k \equiv \left(\omega, \vec{k}\right)$ is for high frequencies $\omega > \omega_c(x)$ and short wavelength components $\lambda_i < \lambda_{ci}(x), i = 1, 2, 3$. These critical scales at spacetime event x are essentially the reciprocals of the local classical radii of curvature at x that limit the size of the local tangent space over which Local Inertial Frames (LIF) can be defined to good approximation.

$$G_{\mu\nu}(x,k) + \Lambda(x,k)g_{\mu\nu}(x,k) = 8\pi\left(\frac{G}{c^4}\right)T_{\mu\nu}(x,k)$$

(1.3)

where the locally variable quintessent field is for "BIT"[180] Planck area $L_p^2 \equiv \hbar G/c^3$

$$\Lambda(x,k) = \left(\frac{1}{L_p^2}\right)\left(1 - L_p^3 n_s(x,k)\right)$$

(1.4)

Where $n_s(x,k)$ is the smooth coherent nonrandom "superfluid" quantum vacuum number *Wigner phase space density* of Bose-Einstein condensed virtual (off mass shell) electron-positron pairs. Indeed, (1.4) can be written as the "two fluid" model

$$n \equiv n_s(x,k) + n_n(x,k)$$

(1.5)

Where the choppy rough incoherent random normal fluid traditional "zero point" quantum vacuum fluctuations from Heisenberg uncertainty noise *Wigner phase space density* is

$$n_n(x,k) \equiv \frac{\Lambda(x,k)}{L_p}$$

(1.6)

and $n \equiv L_p^{-3}$. The scale-dependent *locally random* zero point macroscopic quantum vacuum fluctuation energy effective mass density is

$$\rho(ZPE) = m_p n_n(x,k) = \sqrt{\frac{\hbar c}{G}}\frac{\Lambda(x,k)}{L_p} = \sqrt{\frac{\hbar c}{G}}\frac{\Lambda(x,k)}{\sqrt{\frac{\hbar G}{c^3}}} = \frac{c^2}{G}\Lambda(x,k)$$

(1.7)

---

[180] "IT FROM BIT" John Archibald Wheeler, here the Bekenstein quantum gravity bit.

The density-pressure zero point fluctuation *equation of state*[181] of the locally normal quantum vacuum fluid is

$$\rho(ZPE) + \frac{p(ZPE)}{c^2} = 0 \tag{1.8}$$

with *active gravity mass density*

$$\rho + \frac{3p}{c^2} \rightarrow -2\rho(ZPE) = -2\frac{c^2}{G}\Lambda(x,k) \tag{1.9}$$

The effective static Poisson equation of general relativity for the gravitational potential energy per unit test mass V in the weak curvature small speed Newtonian-Galilean limit is, in general

$$\nabla^2 V = -4\pi G\left(\rho + \frac{3p}{c^2}\right) \tag{1.10}$$

where a positive active gravity mass density produces a universal attractive gravity field.

In our new special case of the macroscopic quantum vacuum

$$\nabla^2 V(ZPE) = 8\pi c^2 \Lambda(x,k) \tag{1.11}$$

The Wigner phase space density quintessent field $\Lambda(x,k)$ can go negative corresponding the "dark energy" or "missing mass of the universe" with anomalous red shifts. This happens when the superfluid density of virtual electron-positron pairs in the Bose-Einstein condensate of the center of mass motions of the pairs exceeds the reciprocal Planck volume. The opposite region $\Lambda(x,k)$ positive, where the coherent Bose-Einstein condensate density of virtual electron-positron pairs is too low, corresponds to Kip Thorne's "exotic matter" causing universal anti-gravity with anomalous blue shifts[182] needed to keep the wormhole open and free from horizons and singularities. It is also needed for

---

[181] John Peacock's "Cosmological Physics" pp 25-26
[182] Perhaps some gamma ray bursts and super high energy cosmic rays.

Alcubierre's warp drive and to explain why the expansion of the universe is speeding up rather than slowing down. Both of these new forms of energy are not real matter on the mass shell, but a kind of virtual matter, a new macroscopic quantum phase of the physical vacuum. The ordinary vacuum we live in is also macro-quantum, but it is balanced at $\Lambda(x,k) = 0$. We now understand, for the first time in the history of modern physics, how to correctly explain *why* the random positive zero point quantum fluctuation energy from Heisenberg's uncertainty principle does not make a huge anti-gravitating cosmological constant that would not even allow the seemingly classical universe that we observe in astronomy to come into being. This solves the problem Andre Sakharov raised in 1967.

Now we go back to the details of Kip Thorne's toy model proto-star gate metric in the light of my new theory partially and incompletely outlined above. For simplicity, I suppress the additional complexity of the scale dependence. That will be understood. The point is that macroscopic quantum geometrodynamics is a theory for all scales. We need the energy density ρc^2, the radial tension τ, and the *transverse pressure* that we call p_T. [183] From Visser's (2.67) in the Schwarzschild coordinates

$$T_{\tilde{t}\tilde{t}} \equiv \rho c^2, T_{\tilde{r}\tilde{r}} \equiv -\tau, T_{\hat{\theta}\hat{\theta}} = T_{\hat{\phi}\hat{\phi}} = p_T \qquad (1.12)$$

The source term on the RHS of (1.3) is from ordinary real matter on mass shell. We are not interested in that. We are studying the new forms of virtual matter off mass shell inside the macroscopic quantum physical vacuum. Recall from Feynman's quantum electrodynamics that real matter on the mass shell corresponds to the pole of the single particle propagator[184] in the complex energy plane. Virtual matter is all the stuff not on the pole that has physical effects like in the Lamb shift in hydrogen, anomalous magnetic moment of the electron and the Casimir force and the other tests of quantum electrodynamics that agree with experiment to extraordinary accuracy despite the obscure "scandalous shell game"[185] of renormalization. Therefore, let the RHS of (1.3) be set approximately

---

[183] Do not confuse the transverse pressure with the pressure in the equation of state (1.8).

[184] For both the Maxwell electromagnetic and the Dirac electron-positron fields. This can be generalized to standard model of source fermion lepto-quarks with electroweak-strong force bosons.

[185] Private discussion lasting maybe an hour that I had with Feynman in his Cal Tech office in ~1968 about his own work. He said it was a "scandal" that no one had come up with a better idea than his own "shell game".

to zero. Suppressing the k scale dependence that allowed me to use the idea of the Wigner phase space density, the macroscopic quantum geometrodynamic local field equation we are studying in the SSS metric of (1.1) is

$$G_{\mu\nu}(x) \equiv R_{\mu\nu}(x) - \frac{1}{2}R(x)g_{\mu\nu}(x) \approx -\Lambda(x)g_{\mu\nu}(x) \equiv 8\pi\frac{G}{c^4}t_{\mu\nu}(ZPE)$$

(1.13)

There is a very slight formal, but superficial, resemblance to the classical Yilmaz theory here but the physical idea is completely different. Remember, the ordinary vacuum has $\Lambda(x) \to 0$ to a really good approximation in most situations except for the interesting new anomalies of dark matter, gamma ray bursts, super high-energy cosmic rays, anomalous quasar redshifts and the accelerating universe recently observed. Last but not least, we have the persistent serious paradox of zero point energy and the cosmological constant that not even the great Stephen Hawking has a clue about how to explain. This paradox is like the similar one of 1900 that led Planck to introduce the quantum of action h to explain the spectrum of black body radiation.

From (1.1) and (1.12) in the SSS special case

$$t_{\hat{t}\hat{t}}(ZPE) \equiv \rho(ZPE)c^2 = -\frac{c^4}{8\pi G}\Lambda(r)g_{\hat{t}\hat{t}} = \frac{c^4}{8\pi G}\Lambda(r)e^{2\phi(r)}$$

$$t_{\hat{r}\hat{r}}(ZPE) \equiv -\tau(ZPE) = -\frac{c^4}{8\pi G}\Lambda(r)g_{\hat{r}\hat{r}} = -\frac{c^4}{8\pi G}\frac{\Lambda(r)}{\left(1-\frac{b(r)}{r}\right)}$$

$$t_{\hat{\theta}\hat{\theta}}(ZPE) = t_{\hat{\phi}\hat{\phi}}(ZPE) = p_T(ZPE) = -\frac{c^4}{8\pi G}\Lambda(r)$$

(1.14)

Then, from Visser's generic eqs (2.82) to (2.84), the special case of the local partial differential field equation (1.13) for the macroscopic quantum vacuum is equivalent to the following system of three *one-dimensional* radial coupled nonlinear ordinary differential equations

$$\frac{\partial b(r)}{\partial r} = \Lambda(r)e^{2\phi(r)}r^2$$

(1.15)

$$\frac{\partial \phi(r)}{\partial r} = \frac{b(r) - \dfrac{\Lambda(r)r^3}{\left(1 - \dfrac{b(r)}{r}\right)}}{2r^2\left(1 - \dfrac{b(r)}{r}\right)}$$

(1.16)

$$\frac{\partial \tau(r)}{\partial r} = \frac{c^4 \Lambda(r)}{8\pi G}\left[e^{2\phi(r)} - \frac{1}{\left(1 - \dfrac{b(r)}{r}\right)}\left(\frac{\partial \phi(r)}{\partial r} + 2\frac{\left(1 - \dfrac{1}{\left(1 - \dfrac{b(r)}{r}\right)}\right)}{r}\right)\right]$$

(1.17)

These are the Wheeler "IT FROM BIT" type equations since Λ is a BIT field and the geometrodynamic field quantities $b(r), \phi(r), \tau(r)$ are IT fields. Note that (1.15) and (1.16) for the redshift function $\phi(r)$ and the traversable wormhole mouth shape function $b(r)$ can be solved together without (1.17) in the approximation that variable Λ is externally given. In fact is not because there is also a locally curved spacetime covariant spontaneously broken vacuum symmetry Landau-Ginzburg type "BIT FROM IT" field equation with a "wine bottle bottom" (AKA "Mexican Sombrero") effective potential.[186]

---

[186] The massless Goldstone modes are the small phase oscillations around minimum trough of the effective potential for $m^2 < 0, \xi > 0$. The massive "Higgs" modes are the small amplitude oscillations up the sides of the potential.

$$g_{\mu\nu}(x)D^{\mu}D^{\mu}\Psi(x)+m^{2}(x)\Psi(x)+\zeta(x)\left|\Psi(x)\right|^{2}\Psi(x)=0 \tag{1.18}$$

Where

$$\Psi(x)\equiv\sqrt{n_{s}(x)}e^{i\Theta(x)} \tag{1.19}$$

and

$$g_{\mu\nu}(x)\approx\frac{1}{2}L_{p}^{2}\left(\hat{D}_{\mu}\hat{D}_{\nu}+\hat{D}_{\nu}\hat{D}_{\mu}\right)\Theta(x) \tag{1.20}$$

The D^{μ} are spacetime covariant derivatives. The $\hat{D}_{\mu}$ are electromagnetic gauge covariant derivatives in case an external electromagnetic field is switched on. Otherwise they are ordinary partial derivatives. The macroscopic quantum metric tensor geometrodynamic field in (1.18) is the long wave infrared limit of the "World Crystal Lattice" of Hagen Kleinert, where curvature comes from vortex strings in the superfluid vacuum order parameter $\Psi(x)$ that correspond to disclination lattice topological defects. Torsion, not found in Einstein's 1915 GR, but considered by him in later uncompleted versions of "unified field theory", corresponds to vortex strings that are dislocation defects in the World Crystal Lattice. For more details on this program for research see http://stardrive.org/Jack/open.pdf

"The Star Gate Conspiracy" by Jagdish Mann

A review of the book by Picknett & Prince[187]

"Man is an amphibious being", said Aldous Huxley. "Simultaneously living in two contradictory worlds". On the one hand, through his imagination, his visions, his philosophic and mathematical concepts, and more and more, through his scientific discoveries, he has touched the face of eternity. Yet his personal existence is heartbreakingly short, a mere bubble ready to pop. There is no reprieve for anyone because death is an utterly democratic institution. Every one dies: the rich, the poor, the genius and the village idiot, the beautiful Di as certainly as the unnoticed wallflowers. There is no certainty more certain than that of personal death, not even taxes. It is no wonder that Man weaves myths in which he is an heir to some lost Eden and life ever lasting. On the razor-sharp edge of understanding of the idea of "forever" and his own impending death, Man discovers religion, music, poetry, science, and yes, even his wanderlust to reach the stars. In fact, it is at this juncture that Man fully becomes a man, a creature unique in his loneliness. Capacity for compassion and love is born of this terrifying awareness. The dark door of death is also his window to the mysterious, his connection to everything larger than himself. Unfortunately, it is also the area where he is most vulnerable to banal superstitions, empty promises and simplistic panaceas. Each Age brings its own special imagery and language to these superstitions. At one time it was the spirits in the trees, angels on the head of a pin or the ghosts of the dead moving furniture around in Madam Blavatsky's darkened living room. Now it is the aliens in space ships. There is one common element in all these beliefs, however, and that is the passivity and helplessness it ascribes to the human spirit. Someone or something from out there, be it gods or aliens, will come and save us from ourselves. Where there is superstition there are, like maggots in a corpse and just as abundant, charlatans and hucksters writing books, roaming the lecture circuit, and living off the spiritual need and greed of their fellow men. I, however, do not mind them. They may even provide a necessary service by digesting the carrion of superstitious belief systems. Still there is a satisfaction, even a morbid fascination, when these fellows are exposed. The book "The Star Gate Conspiracy" had the opportunity to do this. The authors Lynn Picknett and Clive Prince did a prodigious amount of research into arcane beliefs of the past and present, and have, to the point they tried, been successful in debunking many

---

[187] This book "The Star Gate Conspiracy" spends several pages describing Jack Sarfatti's life and ideas from their point of view.

silly scenarios. The sunken city of Atlantis is given a short shrift, as is the idea that the pyramids were built some twenty thousand years ago by visiting spacemen. Authors have closely studied the occult beliefs of such people as Madam Blavatsky, Edgar Cayce, Aleister Crowley, R.A. Schwaller de Lubicz and many others of similar ilk. But, unfortunately, the book never follows through. There is no discussion of how to tell one thing from another, to separate the wheat from the chaff, the truth from the con. The authors are ambiguous about even their own beliefs concerning these things. The Devil after all, is in the details, as is God. The spirit of scientific discovery is the history of searching for more details, looking for answers to finer and finer distinctions. But in this book, and in spite of all the research, none of this happens. The authors have another, and from my point of view, flawed agenda, an agenda of scaring the reader into believing in some grand conspiracy. The aims of the conspiracy-other than a vague threat of a new hybrid religion sneaked up on the masses- are never spelled out. The parameters of the conspiracy are so amorphous and encompassing that just the cast of conspirators will boggle your mind. It includes- in addition to the already mentioned occultists such as Crowley and Cayce and their followers- the Knights Templars, the Freemasons, the Rosicrucians, the Theosophists, and, of course, MI5 and the CIA, not to mention the Government of Egypt. I am surprised they left out the Catholic Church and the Republican Party. This is not exposing but joining the bottom feeders, only at even a lower level of the food chain. And in the end, no convincing case is made that there "really is a 'star gate conspiracy' eating at the heart of democracy, human autonomy and decency itself". The book finally is just another attempt at fear mongering. But what is even sadder is that they fail to take the opportunity to shed much needed light on this very important and timely subject of the possibility of space travelers, if any, amongst us. It is obvious that they have done a thorough research on the subject and must have read reams of material. The bibliography alone listed at the end of the book is sixteen pages long and the index is another twenty. Reading the book one realizes that they must have contacted via e-mail, telephone, or personally, almost everyone in the world who has anything to say on the subject. Yet, at the end, no clear conclusion is offered.

The possibility that there is sentient life other than humans somewhere in our universe cannot be discounted. In fact it is horrible to imagine that we are alone in this vast and indifferent universe. I want to believe that there are others out there. But belief is not a fact. Belief is the end of discovery, a stop to further search. Why would you still be looking for a black cat of knowledge in a dark room if you think you have already found her?

As to the presence of ET on our planet, I am not only skeptical but also hostile to the idea. My skepticism is based on the fact that other than testimonials of believers, there is no other evidence for such things. My hostility stems from the fact that this sort of belief system tends to breed a cargo cult dependency, a false hope that someone will soon come and fix all our problems. This is emasculation at its worst.

Well, this is no time for spiritual laziness. We alone are responsible for our personal lives as well as our planet. We have to clean up our own mess. We pay a high price for being human and in return the universe is part of our patrimony. Let us not relinquish it to someone out there, real or imaginary. Let us put our own house in order and find our own way to the stars that beckon us.

The Genius And The Golem by Jagdish Mann

"A man who has seen the unicorn is not to be fooled by a pig."

The Dancing Wu Li Masters

I write this is to go on record for what I remember of Gary Zukav and the writing of "The Dancing Wu Li Masters" and the role Dr. Jack Sarfatti played in it.[188] I am in a good place to be a witness to this. I was around from its inception to its finish. In fact, some of the later chapters of the book were written in La Jolla and at my sister's house in El Centro when Csaba Szabo and I had, at Gary's own request, taken him to get him away from North Beach for a month or so.

When Gary first started working on the book, all three of us, Jack, Gary and I lived in North Beach and saw each other almost daily. It was Dr. Sarfatti who tutored him in the intricacies of quantum mechanics. Gary did not know any physics then. His MA was in psychology. There were hours of these tutorial sessions. Many of these took place in the Caffé Trieste and are remembered by many other North Beach residents as well. But I also remember other situations where Gary asked and received technical information from Jack Sarfatti. Gary was nothing if not persistent and Jack nothing if not patient. One incident stands out in my memory Gary was at that time of his life, for all I know, still is, a

---

[188] "Jack Sarfatti, Ph.D., Director of the Physics/Consciousness Research Group, is the catalyst without whom the following people and I would not have met..." Acknowledgments, "The Dancing Wu Li Masters. Gary Zukav reneged on his witnessed oral contract with Jack Sarfatti for 10% of the royalties. Jack's actual role in the writing of the book has been deleted more and more with later editions like the rewriting of history in George Orwell's "1984".

totally single-minded person and brought an obsessive, and not always attractive, tenacity to his task. On this particular evening, Jack and I were both walking to have dinner with Marcia MacLain who lived on the corner of Green and Kearny when Gary found us. He had come with a purpose and right away barraged Jack with questions. After standing there for ten minutes on the street corner, I reminded Jack of the need to go. Gary just tagged along. As Marcia let us in to her second story flat, he climbed the stairs with us still bombarding Jack with his questions. He carried this on with only a perfunctory acknowledgement of our hostess. Marcia looked at me questioningly and I shrugged, so she put another plate on the table. All through the dinner, Gary stuck to his agenda, speaking between bites. After dinner, I escorted both Jack and Gary to the door and stayed behind to privately apologize to Marcia for having brought an uninvited and unappreciative guest to her carefully planned cooking. The last I saw of them from the second story window was them walking down Kearny Street, engrossed in an animate discussion.

There is another important contribution that Jack Sarfatti made to the writing of "The Dancing Wu Li Masters." He not only tutored Gary enough in physics so that he could ask intelligent questions from other physicists, he even introduced him to people like Dr. Henry Stapp, and took him to Esalen to meet many more.

Seat of the Soul and Oprah Winfrey[189]

I cannot say that without Jack Sarfatti, Gary Zukav could not have written "The Dancing Wu Li Masters, but I can say with certainty, that without Jack Sarfatti, it would not have been the same book. It would have been more like Gary's recent work, "The Seat of the Soul," a book full of trite truths and platitudes held together with Talk Show spit.

Dancing Wu Li Masters - Jack Sarfatti = Seat of the Soul ☺

---

[189] New York Times, October 29, 2001 " 'Oprah' Gaffe by Jonathan Franzen Draws Ire and Sales…In 1996 Jonathan Franzen wrote an essay in Harper's magazine about…the banal ascendancy of television…Last selection but politely withdrew the invitation to appear on her show. And instead of rallying to Mr. Franzen, most of the literary world took her side, deriding him as arrogant and ungrateful…Mr. Moody said…'If you want to sell 700,000 copies," he said, "then you have to play ball with the 700,000-copy vehicles, and then you are in Oprah-land." I am reminded here of Howard Roark's courtroom speech at the end of Ayn Rand's "Fountainhead" with Oprah as Ellsworth Toohey, Gary Zukav as Peter Keating and me, of course as Howard Roark. E. Annie Proulx author of "Shipping News" has also been critical of Oprah's bid to be Dictator of the Literary and Spiritual Arts. ☺ - Sarfatti comment

On the Hidden Origins of Reagan's SDI by Kim Burrafato

"I do not know how President Reagan arrived at his decision…" Edward Teller [190]

Jack Sarfatti and Kim Burrafato in 1979 at Savoy Tivoli

---

[190] p. 526 "Memoirs" Ch. 40 "Strategic Defense" (Initiative, "SDI") 2001 Perseus

The following is a narrative of some of the events surrounding Jack Sarfatti's seemingly significant direct input into the formulation of Ronald Reagan's Strategic Defense Initiative that complemented Edward Teller's. Teller, of course, was the key. The world was a dangerous place in 1979. Afghanistan was about to be invaded, the Iran hostage crisis was looming on the immediate horizon, and the Cold War arms race was in full swing, with the Soviets threatening to take the initiative in Europe, as well as in the overall strategic nuclear theater. I was Jack Sarfatti's roommate at the time, in the movable feast[191] that was San Francisco's North Beach.[192]

---

[191] This group included P. Siegel and Michael Weiner who is now Michael Savage on the radio.

[192] "In the half-century that North Beach has supplied a haven for waves of brilliant and erratic thinkers, few have rivaled Jack Sarfatti for comprehensive weirdness an what just might be original genius." by Jerry Carroll in "Another Eccentric Genius in North Beach?" San Francisco Chronicle, May 11, 1981, p.23

Stephen Schwartz on MSNBC Dateline's "Roots of Rage" Pearl Harbor Program of Dec 7, 2001 on 911.[193]

http://www.msnbc.com/m/v/video_news.asp
is the website to see Schwartz in action on prime time TV his segment is "Roots of Rage" Part III "The Saudis" Schwartz wrote about Jack in the San Francisco Chronicle on August 17, 1997

"I'A TOLD YA SO!" THE SCHWARTZ

(Stage center)

---

[193] Schwartz, an expert on "The Venona File" is now writing "The Two Faces of Islam". He is author of "The Strange Silence" on the Contras in Nicaragua, "Intellectuals and Assassins", "From West to East: California and the Making of the American Mind". Schwartz also writes for the Wall Street Journal, New York Post, American Spectator, Commentary, National Review, Jewish Forward etc. For Schwartz's dispute with his ex-boss at Voice of America over their media handling of Al Qaeda see William Safire Op-Ed New York Times, July 1, 2002 and "State Department Outrage: The Firing Of Stephen Schwartz" By Ronald Radosh, Front PageMagazine.com/ July 2, 2002 http://www.frontpagemag.com/Articles/ReadArticle.asp?=1610
Schwartz worked for Bush's Secretary of Defense, Donald Rumsfeld, at ICS in the mid 80's.

http://www.ladyofthecake.com/mel/space/sbimages.htm

Jack Sarfatti and Stephen Schwartz at Caffe Trieste in early 90's.

Close Encounters

Jack was busily working on a fundamental reformulation of quantum theory that would both explain and include such sacrilegious elements as consciousness, gravity, and superluminal information transfer. One of the more provocative physical implications that emerged from his work was the possibility of interfering with ordinary electromagnetic energy propagation based systems inside spacetime using quantum action at a distance outside of spacetime.[194] This

---

"In the ongoing struggle between the Owell soft-liners and the Rumsfeld-Cheney-Wolfowitz hardliners in the Bush Administration, a small outrage has taken place in the realm of the Department of State. Stephen Schwartz, one of the most prominent commentators on the war against terrorism, and particularly on the role of the Saudis, has been dismissed from his post as an editorial writer, assigned to the new Middle East radio network at the Voice Of America... And Robert Conquest, A man who is undoubtedly the most eminent historian of the Communist experience, calls Stephen Schwartz "one of our most determined and successful investigators of the whole Communist phenomenon"

Stephen Schwartz wrote to me on July 4, 2002:

"I was fired from VOA because of the following:
1. I disagreed strongly with the politically correct atmosphere in the so called "Newsroom" that promoted an adversarial position toward the war on terrorism, i.e. an attitude of skepticism about the motives, means, probability of success, fully supported the Bush administration's commitment to defeating Islamofascism.

2. I objected to the plantation mentality present in the workplace, where there is no training, no break-in, no consistency of assignments or standards, compulsory overnight shifts, no lunch period, and none of the collegial give and take present in the normal cultural of journalism.

3. Other, obscure factors have emerged in this unfortunate mess. VOA's tactics have been extremely ugly, really shockingly so, including inquiries into my religious affiliations and opinions.

Stephen Schwartz"

[194] Jack's letter to MIT Technology Review published in 1976 and CIA Memorandum for the Record on Jack's ideas in 1979. See also "Pentagon Is Said to Focus on ESP For Wartime Use", William J. Broad, New York Science Times, January 10, 1984.

leads to subject of UFOs, which has interested me since I was a child, almost to the point of becoming an obsession. One of the many peculiar physical effects described in numerous UFO close encounter cases was the interference with electromagnetic systems such as aircraft avionics, automobile ignition systems, and household electrical circuits. UFOs were described as being able to shut such systems down at will. Many of these well-documented reports were from reliable military and civilian witnesses. The movie "Close Encounters" has burned that idea vividly into the public's consciousness. That got us thinking: what if one could design a weapon that could neutralize electromagnetic energy based systems along the lines of UFOs—selectively, and at a distance? We knew about the damaging effects of high-energy electromagnetic pulses on electrical systems, from research into hardening command, control, and communications systems against nuclear attack. Such enormous pulses will cripple or destroy any electronic circuitry exposed to them. But according to numerous witnesses' reports, the electronic systems shut down by UFOs are not fried. They simply turn back on like a light switch being flipped on. Obviously, the ETs are employing some exotic new physics. If we could understand the physics involved in such cases of remote electromagnetic interference, then it would be a straightforward matter to engineer a revolutionary offensive and defensive weapons system. There were some tantalizing clues to how such an exotic physics might work. One of those clues arose out of the now famous Einstein-Podolsky-Rosen (EPR) thought experiment. The EPR experiment was devised by Einstein to illustrate the bizarre paradoxes that could seemingly arise out of the quantum theory, when its physical implications were taken literally.

Faster Than The Speeding Photon

Einstein always considered the quantum theory to be an incomplete description of reality and the EPR thought experiment part of his great debate with Niels Bohr. Einstein, Podolsky, and Rosen showed that, if quantum theory was a complete description of individual particles, it's possible to design an experimental arrangement in which a new kind of quantum information or "qubits" and "e-bits" can conceivably get around in ways faster than a signal limited to the speed of light would permit. Indeed, Einstein did not like this consequence of quantum theory. It took another 35 years or so before an actual experiment was conducted to test this idea. Quantum theory was correct in this conclusion if one did not believe in parallel quantum universes.[195]

---

[195] These parallel quantum universes, as in David Deutsch's "The Fabric of Reality" are not the same as, and are not to be confused with, the many classical material brane universes next door to ours in the hyperspace of Super Cosmos—Sarfatti note.

Background information may be found at

http://www.treasure-troves.com/physics/Einstein-Podolsky-
RosenParadox.html
http://members.ozemail.com.au/~kglaszio/page34.html
http://citeseer.nj.nec.com/context/39995/0

Jack's idea that I helped to trigger with my insistent questions over many hours at the Caffe Trieste in the late 70's was the globally self-consistent "loop in time". A New York Science Times article from the Wigner Conference of the New York Academy of Sciences in 1986 cites Jack on "time loops".[196] However, back in 1979 Jack was unaware of similar ideas of Igor Novikov and Kip Thorne.[197] So the idea of the UFO weapon was the "paradox effect". Attempt to set up a time travel to the past paradox with the device you want to temporarily disable in the time loop. Jack later published this in September 1991 Physics Essays (University of Toronto) causing Lyle Fuller to come up with a very ingenious gedanken experiment in the manner of the Bohr-Einstein dialogues. The problem however with all this is that it violates the statistical structure of quantum theory. Although one can encode a message nonlocally, one cannot locally decode it at the receiver faster than the speed of light and even backward in time from the future. Even modern "quantum teleportation" of qubit information needs a light speed limited classical c-bit signal to make it work. Jack maintains that one needs a new post-quantum theory to do it. Quantum theory would then be a limiting case of this more general theory. Jack has made progress in this direction but that is another story that I do not have time for here.

From Russia With Love

The frequency and magnitude of the ideas escalated until we were soon brainstorming about constructing a comprehensive umbrella over the US and its allies that would effectively shut down the guidance and triggering systems of any incoming ICBMs and their MIRV warheads entering its range. It didn't stop

---

[196] This conference was chaired by the late Heinz Pagels who was a close friend to both Ira Einhorn and Nick Herbert. Heinz strongly attacked Jack's arguments for precognitive remote viewing in a phone call with David Gladstone. Oddly enough, Heinz died falling off a mountain near Aspen Physics Center. He had dreamed his death almost exactly and one finds it at the end of his book "The Cosmic Code". Here is another tragic example of The Destiny Matrix we are all caught in.

[197] Which may have come later after 1979? Saul-Paul Sirag traces Jack's ideas here back to at least 1973. Jack's idea, of course, goes back to his strange phone call(s) in 1953.

there. We thought: What a great way to end the Cold War and establish a mutually designed, built, and administered defensive system between the US and the USSR, and other willing global participants. Jack received a friendly postcard from one Professor Igor Akchurin, around Christmas of 1979. Akchurin was then a theoretical physicist of some repute from the prestigious Soviet Academy of Sciences in Moscow. Somehow, Akchurin had been included in Jack's extensive and often used mailing list. The message on the card was brief and cordial—something along the lines of Happy Holidays and Best Wishes "in the spirit of the New Physics." I remember that quoted phrase vividly. All hell broke loose after that. Jack immediately (and wisely) took that as a sign from Heaven. I don't know if it was Jack or me that suggested we write a letter to Akchurin, detailing our ideas about an ultra-high tech, mutually defensive ABM system.[198] We even got into details about how once we (the US and USSR agreed to embark down that momentous and risky path, we could then start building down (dismantling) our offensive MAD-based missile systems—to everyone on Earth's advantage. We even had the foresight to include mention of the future potential threat of rogue states getting their hands on nuclear weapons through proliferation, as well as terrorist groups. But there was an added dimension to this mutual defense system: the extraterrestrial one. What if some of the ETs flying around Earth's skies are hostile? Just because any ET species capable of visiting this planet would be far more technologically advanced than us, doesn't mean they would necessarily be ethically or morally advanced. How might the planet defend itself, if faced with such a hostile ET threat?[199] The kind of defensive system we were proposing might, if perfected, give us a fighting chance, or at least make it less attractive or likely for an alien species to forcibly interfere here. We included that element in the Akchurin letter, too. In other words, we had laid out Ronald Reagan's, and Douglas MacArthur's[200] (albeit

---

[198] As we go to press, President Bush has informed President Putin that America will proceed with anti-missile defense.

[199] Curiously Francis Ford Coppola produces "First Wave" on the Sci Fi Channel that deals with this scenario in fictional form. Of course this is a popular theme and I do not imply any direct causal relation.

[200] Colonel Phillip J. Corso, who wrote "The Day After Roswell" worked for MacArthur. MacArthur's speech, "Duty, Honor, Country" ends with a reference to ET War in space and also "harnessing the cosmic energy". Jack, in a recent meeting (2002) with a high-ranking USG Intelligence Officer was told that in fact Edward Teller did think of SDI as also a defense against possibly hostile ET aliens. This really surprised Jack because Jack heard Teller say that UFOs were not real at public talks at functions of Rabbi Pincus Lipner's Hebrew Academy of San Francisco in the early 90's where Jack was teaching. Teller was on the Board. There is also an unconfirmed rumor on the Net that CIA's James Jesus Angleton was also very fearful of ET invasion. One of people making this allegation claims to be Angleton's grandson. The problem used to be it ET were here,

prescient) entire SDI vision in a single letter sent to a top physicist in the most prestigious academic science institution in Russia. Of course, we knew that this letter would be intercepted and read by our people before going to the USSR. But that was our intention. What better ways to let those in positions of responsibility know what was really going on? It's easier getting a letter in the hands of a high level DOD or CIA Science Technology analyst that way, than it is writing to them directly. As expected, we never received a reply from Akchurin. Not to worry. The seeds had been planted.

"Spurn Not The Nobly Born"[201]

A few months later a friend of mine, Leila Minturn Dwight[202], showed up in North Beach.[203] The Savoy Tivoli was the happening place at the time. It seemed like almost everyone who was anyone hung out there. We introduced the attractive young niece of deceased Andy Warhol diva Edie Sedgwick to the Savoy crowd. Leila fit right in. Not long after that, Leila introduced us to A. Lawrence Chickering. "Lawry" Chickering was heading up a newly formed neo-conservative think tank based in San Francisco at the time. The Institute for Contemporary Studies was set up by soon to be Reagan cabinet members Ed Meese and Cap Weinberger, and others, to study key cultural and political issues of the time. [204]

One of the first big issues that came up in conversations with Lawry was the significance of the New Physics to everything from geo-politics to religion. Chickering[205] had read Gary Zukav's[206] recently published New Age book on the

---

how would he/she/it get here? Now with the parallel universes of Super Cosmos and the kind of star gate warp drive time travel physics Jack is working on, that seems less of a problem. Of course, this is still speculative and conjectural and Jack is first to admit he could be wrong on all of it.

[201] Iolanthe, Gilbert & Sullivan

[202] Descended from Timothy Dwight, first President of Yale. Lawry's grandfather and Leila's great grandfather, Henry de Forest, ran the Southern Pacific Railroad with E.H. Harriman including a private railway car now in the Southern Pacific Railway Museum in Oakland, California. - Sarfatti comment

[203] San Francisco

[204] Jack introduced Stephen Schwartz to Lawry Chickering. Chickering hired Schwartz at ICS.

[205] Lawry was also close to Esalen's Michael Murphy and Jack's old Cornell teacher, Herb Gold.

[206] Zukav has since attracted a huge following because he is being promoted by Oprah Winfrey. Jack, arms flailing about wildly laughingly aping Mel Brooks, calls Gary Zukav "Oy vey Jose, my Golem, My Frankenstein Monster run amuck!"

development of the quantum theory, "The Dancing Wu Li Masters," and had, consequently, developed a keen interest in physics and its relationship to philosophy and theology. Since Jack was instrumental in the writing of the physics parts that book[207], his words carried even more weight with Chickering. Jack spent hours with Lawry, discussing the cultural significance of his emerging "post-quantum" theoretical worldview. Chickering made it known to us on more than one occasion that he had the ears of Republican presidential candidate Ronald Reagan's potential cabinet members and advisors. A number of Jack's "Memoranda for the Record" were later passed on to key people[208] in the newly elected Reagan Administration. Coincidentally, around the same time, Jack was also introduced by telephone to Cap Weinberger's son, Cap Jr. through Joe Lynch, a mutual friend of ours. Although Jack and Cap junior never met face to face, Cap Jr. quickly became a fan of Jack's, and stated he would also forward some of Jack's material to his father.[209] We were flabbergasted when Ronald Reagan gave his first major foreign policy speech of his new administration on March 23, 1983. His now famous "Star Wars" speech, as the then openly anti-Reagan liberal press condescendingly termed it, blew our minds when we read the transcripts.[210] It was as though he'd taken all of our ideas and elegantly distilled them down to a version that the American public could easily understand.

"Never has there been a more exciting time to be alive — a time
of rousing wonder and heroic achievement. As they said in the

---

[207] Not the New Age jargon in it, which was Gary's original contribution like in "Seat of the Soul".

[208] Especially Paul Nitze.

[209] Which he did. Jack received a letter from Cap Weinberger Sr. on DOD stationary thanking him for the material. Cap Jr. said that Jack's ideas were discussed explicitly between Reagan and Cap, Sr. in the White House. Jack presented the detailed ideas of Hollywood billionaire, the late Marshall Naify. Marshall presented a detailed "Star Wars" plan in a crucial lunch at Enrico's in 1981 with Jack and Lawry Chickering. Reagan, of course, knew who Marshall was, whether Reagan was aware of this connection we do not know. It would be an interesting topic for a historian. Marshall financed the James Bond movies with Al Brocoli. See also "Meteor" with Sean Connery. Indeed Marshall had one of the black Aston-Martin convertibles from the Bond films. Jack's report on Marshall's Vision went to Paul Nitze.

[210] Reagan also specifically alludes to Jack in his 1986 State of the Union Address where he says physicists have found evidence for the existence of God in their equations and he mentions "Back to the Future". This was two years before Hawking's questions in a "Brief History of Time". The origins of these remarks can be traced to his discussions with Cap Weinberger, Sr. on Jack's ideas.

film, 'Back to the Future': 'Where we're going, we don't need roads.' Well today, physicists peering into the infinitely small realms of subatomic particles find reaffirmations of religious faith: astronomers build a space telescope that can see to the edge of the universe and possibly back to the moment of creation...America met one historic challenge and went to the moon. Now America must meet another — to make our strategic defense real for all citizens of Planet Earth...The American dream is a song of hope that rings through the night winter air. Vivid, tender music that warms our hearts when the least among us aspire to the greatest things — to venture a daring enterprise; to unearth new beauty in music, literature and art; to discover a new universe inside a tiny silicon chip or a single human cell."
Ronald Reagan, State of the Union Address, 1986

Jack Sarfatti on CV 61, USS Ranger, Indian Ocean, 1987

Kim Burrafato meets Richard Nixon. Kim explains Sarfatti's physics and its relevance to the Strategic Defense Initiative.

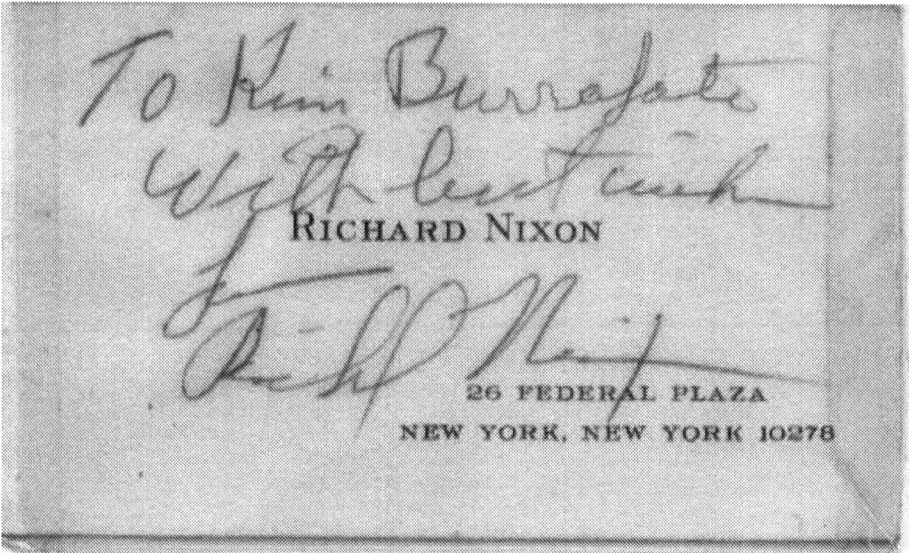

THE SECRETARY OF DEFENSE
WASHINGTON, THE DISTRICT OF COLUMBIA

1 0 AUG 1984

Dr. Jack Sarfatti
POB 26548
San Francisco, California   94126

Dear Dr. Sarfatti:

   Thank you for the information you
sent me on your work on Superluminal Group
Physics. I have provided the information
you sent me to my technical staff. I wish
you luck with your future experimentation.

                    Sincerely,

                    Casparld W. Weinberger

The Famous Chickering Letter [211]

---

[211] To Richard DeLauer at DOD. Note reference to "untappable unjammable command control communications" years before IBM's Charles Bennett et-al introduced "quantum cryptography", "dense coding" and "teleportation". Chickering's letter is more important today in 2002 because of Jack's argument below that the "no-cloning theorem" of quantum computers may not be correct. This would mean stand alone nonlocal quantum C^3 that does not need a classical signal link. Chickering's letter was inaccurately described by Jeremy Bernstein in The New Yorker article "Quantum Engineers" on John S. Bell, by Murray Gell-Mann in "The Quark and the Jaguar" on "Flap Doodle" and by N. D. Mermin in Physics Today. Note reference to "The Dancing Wu Li Masters" and CIA Memorandum for the Record that Lawry had a copy of. Gary Zukav removed much of the final chapter on Jack's ideas in later editions and did not pay Jack the 10% of royalties to the book that he had promised and agreed to in front of several witnesses including myself. "One physicist at the New York Meeting, Dr. Jack Sarfatti of San Francisco said that he not only believes that faster than light communication is possible by means of 'time loops,' but that he is trying to attract backing from the Defense Department..." Malcolm W. Browne, "Quantum Theory: Disturbing Questions Remain Unresolved", New York Times, Feb. 11, 1986, p. 25. See also NOVA and Ultra Science TV shows on "Time Travel" citing an experiment in Germany in which Mozart's 40th Symphony allegedly is transmitted faster than light by quantum tunneling of photons through a barrier. Ray Chiao of UC Berkeley disputes this however. Jack met William F. Buckley, Jr. at Chickering's house around the time Buckley was debating Carl Sagan on "nuclear winter". Buckley asked Jack what he thought of the issue. Jack had first met Buckley with Norman Thomas in a church in Manhattan in 1956 as a Cornell Freshman. Jack also spent several hours sitting with Itzak Rabin and Edward Teller at the home of Rabbi Pincus Lipner when Jack was teaching at the Hebrew Academy of San Francisco. Jack had met the prominent San Francisco businessman, Alvin Duskin, also ~ 1980 and told him about prospects for zero point energy. Jack ran into Alvin only recently at the San Francisco Bay Club. Alvin told Jack that his meetings with him 20 years earlier significantly influenced him to get into the alternative energy business (flywheels from Lawrence Livermore Laboratory). Curiously, Alvin and Lawry Chickering were working together on the "Educate the Girls Foundation" for Muslim women. Alvin and Lawry visited Jack and Creon Levit at ISSO in the Dakin Building at 3220 Sacramento Street to look at the possibility of sharing office space. They wound up at the Presidio.

March 12, 1982

Richard D. DeLauer
Under Secretary of Defense for Research and Engineering
Room 3E1006
The Pentagon
Washington, D. C. 20301

INSTITUTE
FOR

CONTEMPORARY
STUDIES

Dear Dr. DeLauer:

I am writing to say how much I enjoyed sitting next to you at
the Secretary's dinner last week (March 2) and to follow up on
our conversation about the work of a physicist friend of mine
which just might have profound implications for certain aspects
of the technology of warfare.

My friend's name is Dr. Jack Sarfatti. He has worked with
physics Nobel Laureate Abdus Salam, at the IAEA-UNESCO International Centre for Theoretical Physics in Trieste, the United
Kingdom Atomic Energy Research Establishment at Harwell, Birkbeck College of the University of London, and was on the physics
faculty of San Diego State University. Jack graduated from
Cornell in 1960 and received his Ph.D. in theoretical physics
from Berkeley. He feels his current work may have important
application to a subject emphasized by Cap at the dinner--name-
ly, problems of command, control, and communications with our
submarine force.

I am writing to you as a layman, with only a layman's knowledge
of these issues. I came to know Jack after reading Bantam's
The Dancing Wu Li Masters, by Gary Zukav. The award-winning
book is an exposition of the "new physics," i.e. the physics of
quantum mechanics, relativity, and how to put them together
properly. The final chapter describes an early version of
Jack's ideas, which have since been formalized to some extent.
More recently the CIA (Memorandum for the Record, December 4,
1979) described Jack's intuitive ideas as highly speculative
but "genuine basic research" lacking experimental support.

Jack says this is due to lack of funds; but he believes that the
series of experiments now completed, the latest by Aspect at the
Instituto of Optics in Orsay, does in fact provide enough
evidence to justify going further with the effort. These experiments suggest a "nonlocal" action-at-a-distance between
spacially separated but quantum correlated fragments from a

Page two: DeLauer

decaying quantum system. Each fragment "senses" what is happening
to the other in spite of the fact that no ordinary electromagnetic
or acoustic signal is connecting them (e.g. pairs of photons or
electrons on fragments in "molecular dissociation" chemical reactions)

If this sounds like occult "sympathetic magic," it appears to be
an essential feature of the quantum theory that Einstein discovered
in 1935 much to his dismay with his students Podolsky and Rosen.
Many good physicists have given "proofs" that this nonlocal faster-
than-light effect exists but cannot, in principle, be controlled to
send useful messages for a command-control-communication application. Jack and a small group of other "maverick" physicists think
otherwise and also think that the "proofs" make an assumption that   --
may be unjustified.

Jack says that if in fact we can control the faster-than-light nonlocal effect, it would be possible, using something like the "
"molecular dissociation" approach, to make an untappable and unjammable command-control-communication system at very high bit-rates
for use in the submarine fleet. The important point is that since
there is no ordinary electromagnetic or acoustic signal linking
the encoder with the decoder in such a hypothetical system, there
is nothing for the enemy to tap or jam. The enemy would have to
have actual possession of the "black box" decoder to intercept the
message, whose reliability would not depend on separation from the
encoder nor on ocean or weather conditions!

I know it sounds like science fiction, but no one honestly knows
for sure at this point. Jack has repeatedly emphasized that he
could be dead wrong in his hunch about this. He says it is important to know for sure whether or not such an application of his
far-out thinking is or is not possible. My sense is that testing
his theories would not be either difficult or costly, and God.
knows what other clever ideas he and his crew of eccentric geniuses
might come up with if they were properly supported?

Pierre Noyes, a physicist at Stanford Linear Accelerator Center,
recently told Jack that there is a network of European physicists,
centered in Italy, who are actively trying to design practical
faster-than-light quantum communicators. Jack suspects that Igor
Akchurin, a member of the Soviet Academy of Sciences and Director of the Moscow Institute of Philosophy, is thinking along
lines similar to his own. (I understand that Akchurin even sent
Jack a Christmas card a couple of years ago with an inscription
like "Yours in the telepathic spirit of new physics.")

Jack has been playing with several "Gedankenexperiments" that
attempt to control the faster-than-light effect. He is in touch
with a small number of other physicists who are thinking about
such things. He also now believes that he has the seeds of a

Jack Sarfatti

rigorous mathematical model that extends the structure of Einstein's relativity from a "real" continuum to what he calls a "quaternion" continuum. This extension contains, he says, the nonlocal effect in a beautiful way and also gives the unified electromagnetic-weak-nuclear forces model for which Salam shared the Nobel Prize in 1979.

I have seen two positive letters of support to Jack from Salam in the past couple of years, and John Wheeler, who developed the theory of nuclear fission with Bohr, has been replying to Jack's prolific output of ideas with friendly cards and copies of his own work on "delayed choices" Gedankenexperiments that are closely allied to Jack's ideas. Wheeler does not, however, go as far as Jack on the control issue. In fact, he once described Jack's attempt to control this effect as "moonshine." On the other hand, Jack is fond of telling the story of how Lord Rutherford, discoverer of the atomic nucleus, in the thirties used the same word in relation to the idea of a nuclear bomb.

I think I have written enough. I do hope that you or one of your associates will be sufficiently interested in Jack's work to contact him. You may reach him either through me, or c/o P. O. Box 26548, San Francisco, CA 94126. I am enclosing for your interest some articles from popular journals on the issues considered in this letter--from Science Digest, The Intelligent Machines Journal, and The Economist.

Again, I appreciate your interest and hope something comes from the contact. Please thank Cap again for me for an extremely interesting and stimulating evening.

                                    Sincerely,

OFFICE OF THE UNDER SECRETARY OF DEFENSE

WASHINGTON D C 20301

RESEARCH AND
ENGINEERING
(R&AT)

12 JUL 1984

Dr. A. Lawrence Chickering
Executive Director
Institute for Contemporary Studies
260 California St
San Francisco, CA 94111

Dear Dr. Chickering:

This is an interim response to your letter dated 12 March 1982 to Dr. DeLauer regarding the work of Dr. Jack Sarfatti, which has been recently forwarded to our office.

Dr. Sarfatti's speculations regarding the possible application of "non local action-at-a-distance" theories to secure communication systems could be of interest to the Department of Defense if verifiable. Scientists have speculated over the possibility of "faster than light" communications since Bell's famous experiment in 1965. A summarization of Dr. Sarfatti's latest findings and conclusions along these lines would aid our determination of the potential benefits offered by his research and in arranging for any appropriate further discussions.

Thank you very much for your interest in the research and development programs of the Department of Defense.

Sincerely,

Domenic A. Maio
Colonel, USAF
Military Assistant to DUSD(R&AT)

cc:
Col Kenneth Hollander

151

EMBASSY OF THE
UNITED STATES OF AMERICA
Paris, France

May 7, 1982

Mr. Jack Sarfatti
W) 26548
San Francisco, Calif.

Dear Jack,

Just a quick note to thank you for your long letter and the "dossier" for TV presentation. I am planning to see Alain Aspect after the Reagan visit here and will pass on to him whatever information you wish. I would like to listen carefully to the tape that Lawry and I made of our visit with Mr. Aspect before attempting to describe in what way he feels that Gary Zukav misrepresented him. He does not consider his experiment a test of faster than light phenomena, no matter what the result. He was on his way to a conference in Italy where, I believe, they were going to discuss experiments such as his which would also include faster than light interpretation.

I hope I will have time in mid June to concentrate a little bit on all of this. I am still settling in here as Lawry will tell you. I am in constant touch with Gay Obolensky who thinks that you and he are working on the same wave length, the difference being that Gay has made his discoveries through his inventions and he is able, he believes, to send signals (information) faster than light and can demonstrate this phenomenon (once he gets about $150,000 to set up a demonstration properly). I have sent him a copy of your letter to me as well as the "dossier". His address is:

124 Rte 17
Sloatsburg, N.Y. 10974

and his telephone number: (914) 753-2781

"Letter from Bootsie Galbraith, wife of Reagan's Ambassador to France, to Jack Sarfatti, 1982"

I will be in touch again soon. I do not
suppose there is any chance that you might
be around New York this summer ? I am going
to be with my family in Cornwall, N.Y., which
is near Guy, and it would be great if we
could all get together then.

Meantime, all my best regards.

Brigitte Galbraith

Mrs. Evan G. Galbraith

"Aye, Aye, Captain, Anchors Away!"

Review of Herb Gold's book, "Bohemia: Where Art, Angst, Love, and Strong Coffee Meet", by Jack Sarfatti

(Simon & Schuster, 1993, ISBN 0-671-76781-X)

Herb Gold, Jagdish Mann, Jack Sarfatti, North Beach, 1999

"I love this book! It is a fascinating and hilarious magical mystery tour of that unmapped land called Bohemia writes Pat Conroy, author of "The Prince of Tides".

Jagdish, Lawrence Ferlinghetti, Jindan and Jack at Enrico's

Jack, Enrico Banducci and Jagdish (2000 AD) in fond memory of the many good times with Marshall Naify "who, partnered with Al Brocoli, was a key financier of Ian Fleming's James Bond movies starring Sean Connery."

John Updike

Herbert Gold is a distinguished author and journalist in San Francisco. He has just published a history of Bohemia that will appeal to both cultures literary and scientific. Gold has joined company with John Updike whose books; "The Witches of Eastwick" and "Roger's Version" explain the "New Physics" better to the public than Hawking's editors did in the over-hyped "Brief History of Time".

"Sukie steered the conversation away from herself...Van Horne talked about himself, his hopes of finding a loophole in the second law of thermodynamics[212]. 'There has to be one,' he said, beginning to sweat and wipe his lips in excitement, 'and it's the same fucking loophole whereby everything crossed over from nonbeing. It's the singularity at the bottom of the Big Bang...And what kind of a force is it that operates across space instantaneously and has nothing to do with the electromagnetic field?...There's a formula out there, and it's going to be as elegant as good old $E = mc^2$. The sword from the stone, you know what I mean?...Sukie laughed...Why couldn't a wild man like Darryl blunder into one of the universe's secrets?" pp. 134-5 "The Witches of Eastwick"[213]

---

[212] Updike appears to allude to my work on coupling a hot negative quantum temperature heat reservoir to a cold positive classical temperature reservoir in a Carnot Engine. One then gets complete conversion of heat to work from both reservoirs in a quantum loophole to the second law of thermodynamics. Updike certainly knows the Sedgwicks alluded to in "Eastwick". He also has a lot of things to say about superconductors in this passage in the full version.

[213] Van Horne played by Jack Nicholson. Sukie played by Michelle Pfeiffer in the movie.

"Cartoon of Jack Sarfatti lecturing in Brazil by Suky Sedgwick, 1984"

It takes two to tango!

"Darryl Van Horne" and "Sukie laughed"

There's a formula out there, and it's going to be as elegant as good old $E = mc^2$.

IT FROM BIT + BIT FROM IT = SELF-EXCITED UNIVERSE

$$R_{\mu\nu}\left(herenow\right) - \frac{1}{2}\left[R\left(herenow\right) - \frac{2}{L_p^2}\left(1 - L_p^3 \left|\Psi\left(herenow\right)\right|_{future|past}^2\right)\right] g_{\mu\nu}\left(here - now\right) = -\frac{L_p^2}{\hbar c} T_{\mu\nu}\left(herenow\right)$$

Vladimir Nabokov

I first met Herb Gold when I was an 18-year-old physics major at Cornell.

***Student Princes: Sergio Sismondo[214], David Green (3rd from
left) and Jack Sarfatti (far right) Cornell***

http://www.theatrehistory.com/american/musical024.html

Tom Pynchon was in some of my classes.[215] Gold was filling in for Vladimir
Nabokov who was off writing the screenplay for "Lolita". Herb introduced me to
European literature (Dostoyevski, Goethe's Faust, Tolstoy, Proust). I would see
him outside of class either in Willard Straight Hall or in Noyes Lodge where C.
Michael Curtis[216], Kirk Sale, Richard Farina, Peter Yarrow[217], Tom Pynchon,

---

[214] Italian and Argentinean Steel Family

[215] "Been Down So Long It Looks Like Up To Me" by Richard Farina describes this
group I was part of at Cornell.

[216] Senior Editor of The Atlantic Monthly made important by Ellery Sedgwick in early
20th Century.

Sergio Sismondo and other emerging literati and people of talent would gather. Curiously and uncannily enough, Noyes Lodge was given to Cornell by the grandfather of my far future Grand Passion Suky Sedgwick of six years in honor of her grandmother, Julia Gilman Noyes. Suky's grandfather on her mother's side was Rudyard Kipling's close friend and a partner of Railway Baron, E.H. Harriman. The entire story is to be found in the book, "Edie", by George Plimpton, editor of the Aga Kahn's "Paris Review", and Jean Stein, heiress to the MCA super-fortune left by her father, Jules Stein. Even more remarkable is the fact that her paternal grandfather wrote an essay in 1908 that anticipates the New Physics in which the Future causes the Past. Herb Gold explains all this in Chapter One of his book fitting me into the Bohemia of Jack Kerouac, Gregory Corso and Lawrence Ferlinghetti in that sizzling cauldron of creativity that is San Francisco's North Beach. Gold connects North Beach of the latter 20th century to the Bohemia of Vienna, Paris, London and Greenwich Village of years gone bye. Gold's richness of prose is a joy to read. Gold's book should be made into a docu-drama. Like Mel Brooks, Yiddish Vaudeville would have been for me, were it not dead and were I not partially waspified by the Cornell Savoyards in my tender years when I first met Herb at Telluride House in the Groves of Academe in 1957.

It was in Ithaca with Herb Gold that the imprint of the Penelope archetype was made on my Wandering Jewish Soul that appeared in my future liaisons with several Amazon Women who run with Wolves—and bite like them too!

---

[217] Peter Paul and Mary.

Super Amazon Woman with Jack on her short leash!

I went to Midwood High School in Flatbush in the 50's, as did Woody Allen. I was exposed to some of the same influences that nurtured his comedic genius.[218] My appearance on Gold's stage is in the very beginning (pp.14-17) "Protocols of the Elders of Bohemia". A play on "Protocols of the Elders of Zion" an anti-Semitic tract by the Tsar's Secret Police still loved by New Age UFO conspiracy kooks and crypto-Nazi's of the still-present Occult Third Reich.

http://qedcorp.com/book/psi/hitweapon.html

---

[218] Including Enrico Banducci, though much later.

Morning of the Magicians and The Spear of Destiny

The title of this first chapter evokes memories of pop-occult books like the French "Morning of the Magicians" by Pauwels and Bergier, the "Spear of Destiny"

Look who has it now! ☺

by Trevor Ravenscroft, "The Edge of History"[219] by William Irwin Thompson, and "An End to Ordinary History" by Esalen's Michael Murphy.[220] Gold's prose in this book is very rich with literary images. If one were to check them all out

---

[219] With a chapter on Ira Einhorn at Esalen in Big Sur.
[220] Herb and Michael are close friends. Michael's book is fact described as fiction as is Jacques Vallee's "Fastwalker" with "Slider's" Tracy Torme.

one would get a very good liberal education. The book "Techgnosis" by Erik Davis gives the historical background to put a context on what all this means. Herb writes:

"America may not have been greened, but it was boheemed." p.13 Bohemia is a nonlocal "virtual community" linked by cultural telepathy, the quantum voodoo of Bell's theorem, the sympathetic magick of Aleister Crowley, the Chapel Perilous of Robert Anton Wilson, the occult "morphogenetic fields" of Rupert Sheldrake, the "implicate order" of David Bohm, faster than the confined speeding photons bouncing "higgledy piggledy" in the tiny fiber-optical tubes of that magical "information super-highway" promised us by Clinton, Gore, Gates and Clarke, though seen long ago by Ted Nelson. Bohemia is the highest form of the modern solid Republican neo-conservative idea of the "free market economy", it is the purest expression of democracy, and it is the transubstantiation of Saint Augustine's "City of God" to Planet Earth.

"Like ailanthus, the tree of heaven, Bohemia grows in any alley where there's a bit of fertile dirt and noninterference." p.14

Talking about me, Herb writes:

"The Bohemian physicist...contributes a balanced scientific non establishment for this expanding society. I don't mean to disparage the work; either...among all the blatherers there sometimes appears a breakthrough thinker. Originality has always required a fertile expanse of fumble and mistake. That's the beauty of the option. Your wastrel life might turn out to be just what's required to save the planet." p.14

"Sarfatti's Cave is the name I'll give to the Caffe Trieste in San Francisco, where Jack Sarfatti, Ph.D. in physics, writes his poetry, evokes his mystical, miracle-working ancestors, and has conducted a several-decade-long seminar on the nature of reality and his own love life to a rapt succession of espresso scholars.

LIFE AMONG THE CAPPUCCINO PEOPLE

Cartoon by Norman Quebedeau, Jack standing with hand raised at telephone.
Norman is flying like in a Marc Chagall painting upper right corner.
Next cartoon is also by Quebedeau.
http://caffetrieste.citysearch.com/

He sings Gilbert and Sullivan songs.[221]

Joan Zajac, Jack Sarfatti, Alfred Kahn at Cornell [222]

http://math.boisestate.edu/gas/yeomen/html/yeomen_home.html

---

[221] Photos from high-energy astrophysicist, James Edgar Felten, buddy at Cornell and UCSD. Jim introduced me to Martin Rees in the late 60's at the Institute for Theoretical Astronomy, Cambridge, UK.
[222] Alfred Kahn, Professor of Economics as Sergeant Meryll became CAB head under President Jimmy Carter.

Jack as Colonel Fairfax in white cape in "Yeomen of the Guard"

Cornell 1964

Jack Sarfatti as Alfred. Timothy Jerome as Falke [223], "Fledermaus" 1964, Oberlin Conservatory's Highfield Theater, near Hyannisport, Cape Cod, Massachusetts

http://www.metopera.org/synopses/dieflede.html

---

[223] Tim is in several Woody Allen movies and is in "The Cradle Will Rock" with Susan Sarandon playing Margherita Sarfatti. I was with Roberta Anne Friedman at the time. Roberta taught English for many years at San Diego State much beloved by her students. She died at her desk working apparently of a heart attack. Thanks to James Edgar Felten and Richard Moyer for keeping me informed. Tim will be in the San Francisco production of "La Boheme" at The Curran, Fall 2002.

He suffers tragic reverses among women.

***Jack Sarfatti, Suky Sedgwick and John-Paul Marshall, Buena
Vista Café, San Francisco***

With ample charm and boyish smiles he issues nonnegotiable demands...It's
Jack Sarfatti against the world, and he is indomitable.

One of his soaring theories is that things, which have not happened, yet can
cause events in the present...Obviously this has consequences for prediction, the
nature of causality, our conceptions of logic...He has published papers in
respectable physics journals. His poetry is widely photocopied. His
correspondence with the great in several fields is voluminous, recorded on

computer disks. Cornell University BA, University of California Ph.D., his credentials are impeccable. Following is a quotation from a lecture given to a San Francisco State University physics seminar on 30 April 1991:

Causality-Violating Quantum Action-at-a-Distance?

By Dr. Jack Sarfatti

Jack at the Caffe Trieste

The universe is created by intelligent design but the Designer lives in our far future[224] and has evolved from us [225]...Perhaps all of the works of cultural genius, from the music of Mozart to the physics of Einstein, have their real origin in the future. The genius may be a real psychic channeler whose mind is open to telepathic messages from the future.[226] The genius must be well trained in his or her craft and intellectually disciplined with the integrity of the warrior in order to properly decode the quantum signals from the future. The purpose of our existence would then be to ensure, not only the creation of life on earth, but also

---

[224] Princeton's Richard Gott has a new book "Time Travel" (2001) with essentially this idea years after I suggested it starting around 1973 based on my contact in 1953.

[225] The influence of Harvard's Henry Dwight Sedgwick on my thought here is obvious.

[226] This precognitive remote viewing funded by the CIA and the DIA, as told in James Schnabel's "Remote Viewers: The Secret History of America's Psychic Spies", is a violation of quantum physics but not post-quantum physics. The mathematics of this is in papers by Antony Valentini.

the creation of the big bang itself! We obviously cannot fail since the universe cannot have come into existence without us in this extreme example of Borgesian quantum solipsism. Existentialism is wrong because it is an incorrect extrapolation of the old physics. Breton's surrealism, with its Jungian idea of meaningful coincidence, is closer to the truth. This would then be "The Final Secret of the Illuminati"[227] - that charismatic chain of adepts[228] in quixotic quest of their "Impossible Dream" of the Grail. Enough of my subjective vision, now on to the objective physics." pp. 15-16

I open a page "randomly" (in deference to the hard-boiled skeptics who believe that the Universe is "sound and fury" and that we are accidental) to Gold's chapter "Israel" and on p.115 I find:

"So now I am in the first hour of one of my deaths. The thought made me dizzy. I was reminded of Jack Sarfatti, Ph.D. physicist and reincarnation of the eleventh-century mystic Rabbi Sarfatti...with rapt descriptions of how events from the future cause events in the past."

---

[227] Book by Robert Anton Wilson
[228] Heinz Pagels in "The Cosmic Code" also talks about this as well as his own dream of his death that came true. Usama bin Laden talking of his 911 Attack on America, mentions such precognitive dreams in the horridly evil videotape released by the Pentagon.

Nefertiti's Eye by Jagdish Mann

PROLOGUE

Of the three parts of this story, the Present is true.
The Future is all imaginary but not necessarily untrue.
The Past is for the reader to decide for all eyewitnesses have been long dead.

THE PRESENT

On December 6, 1912, a team of archeologists unearthed the statuette of Nefertiti in Tell el-Amarna. She had lain buried and forgotten in the sand for more than thirty centuries.

This was a great archeological find because until the close of the Nineteenth Century, nothing was known of Akhenaten and Nefertiti, not even their names. This was no accident. Right after Akhenaten's death (some say murder), their names were systematically chiseled off the monuments, their faces were defaced from statues and their new wide boulevard garden city of hundred thousand was razed to the ground, its very bricks stolen and carried off. To this day, it remains an unparalleled example of petty spite and vindictive vandalism.[229]

Still, the attempt did not succeed. In fact, as this story shows, it failed spectacularly and beyond all imaginings.

First, there were the ubiquitous storytellers, poetizing and pantomiming their tall tales, hiding within fiction the facts of a prophetic Pharaoh preaching of One God. And, there were also the stories about a nameless queen. "She was so beautiful, O beloved listener that my tongue is speechless with awe", would say the storytellers. The awe was, at first at least, often aided by the fear of the censor. But like all censorship, this too had the opposite effect. Nefertiti, forbidden to be mentioned, became a legend.

Then with the onset of the modern scientific archeology and increasing successes in deciphering of hieroglyphic and cuneiform writing, these faint whispers began to take shape. Fanatic's hammers were not able to disfigure everything. The border stele in the remote areas had survived the destructive fury of the priests; and the clay-tablet letters written to foreign capitals had also escaped the censor's reach. Archeologists began to read these dispersed messages and started to fill in the empty chiseled-out spaces on the monuments in Thebes and Karnack. By 1900, a picture had emerged of a pharaoh who was courageous or crazy enough to defy one of the most enduring, powerful and entrenched establishments the world has ever seen, that of the priesthood of Amun. He challenged their old orthodoxy and started a new ethic based monotheistic religion and erected a beautiful city 300 miles North of Thebes, away from the old oppressive temples of Amun.

Reaction to this rebellion was swift and complete. Within a year after his death, nothing remained his religion or his city. A concerted and organized effort was made to wipe out even the memory of his reign.

But what about the queen herself was Nefertiti as beautiful as the myth hints and the carved epithets proclaim?

---

[229] Like the Taliban's destruction of the Buddha statues in Afghanistan.

The answer emerged from the sand on that December afternoon with the discovery of the bust of Nefertiti, dug up near the small town of Tell el-Amarna in Egypt. A team of archeologists working for the German Orient Society and led by Professor Ludwig Borchardt of Berlin had been excavating the site. The team was convinced that they had located Akhetaten, the short-lived Capitol City of Akhenaten.[230] They were allowed to dig the site in the name of the Government of Egypt. They were correct in their conclusion; they had found the buried capital, and something more.

The first person to lay eyes on Nefertiti's face in 3300 hundred years was Mohammed Ahmes Es-Senussi. On this day of December, he was digging in room 19-grid p-47. The area was divided in grids measuring 600 square feet and each grid was assigned to one digger. Fates chose Mohammed Ahmes Es-Senussi, a young digger with dancing eyes. As his shovel loosened the impacted sand to slowly cascade downwards, through this waterfall of sand the slanted rays of the afternoon sun lit up the gold and green colors of the queen's necklace. He stood still for a moment, mesmerized. Then he put his shovel gently on the ground and knelt down on his knees and brushed the remaining sand of her face. She had lain buried head down in the debris for all this time. If those eyes could see the young fellahin bending over her on his knees, she would have been pleased. He may have even reminded her of someone she may have once known. And she was used to people kneeling before her. But she could not see him. She was only a statuette, a sandstone figurine barely twenty inches tall, even if she was in near perfect condition. The only visible damage was the chipped earlobes, broken off cobra-head decoration on her crown. Also the inlay of the retina for the left eye was missing. A shout from Mohammed brought all the picks and shovels to a stand still. The great man himself, Professor Borchardt, was sent for from his makeshift hut where he slumbered on a canvas cot after a heavy midday meal and strong German beer. It took another twenty-five years, however, before the discovery was made public. Thievery and deceit was involved in the delay.

Within days, the Germans had smuggled the little bust out of Egypt disguised as broken pieces of pottery. The Egyptian Government found out about the theft soon after from Mohammed es-Senussi, who perhaps was the only person in this whole affair without dishonest motives. The catalyst for Mohammed's complaint, however, was Rasool the water-carrier. He was a spy for a team of French archeologists also working in Egypt.

"Do you know Mohammed, the foreigners have stolen the statue you found and had taken it to the land of the infidels? Now only the fat professor will get credit for it, and much money. But it was not he who found it but you. As Allah is my witness, I saw that with my own two good eyes." Rasool said to

---

[230] The first Camelot.

Mohammed es-Senussi. Mohammed, his curiosity triggered, went to the Germans and demanded to see his find. In a surprisingly incompetent attempt to placate him, they showed him a different statue, trying to pass it for the real thing. Right away, he knew that Rasool was telling the truth. A man who has seen the unicorn is not to be fooled by a pig. Mohammed Ahmes es-Senussi went straight to the Antiquities Department of the Egyptian Government. The German team flatly denied everything and kept their discovery secret from the world and the queen's statuette buried in a Berlin basement for a quarter century more. Still, the German archeologists were not believed and were expelled in disgrace from Egypt, not allowed to return for the next eighty years. The bust, however, was never returned to Egypt and is now on display in lonesome splendor in Das Egyptische Museum in Berlin.

In order to find the missing pieces of the statuette the sand and dirt from room 19 (more than 30 cubic feet) was sifted again and again through a finer and finer mesh. All the ear chips and broken snakehead were finally found but the eye inlay was never recovered. Only later in Berlin, a closer examination revealed that it was never inserted. There are no glue marks in the left eye.

Many theories, some likely and others far-fetched were advanced to explain this unexplainable flaw in this otherwise flawless work. It was suggested, for example, that the artist was interrupted at his work and left the workshop with the inlay in his possession, never to return. Or that the artist had fallen in love with the queen as she posed for him, was jilted by her; and in an impotent revenge, refused to complete his own greatest masterpiece. This is not as farfetched as it may seem at first blush. The queen was known to be free with her favors and capricious with her affections.

Another theory was that Nefertiti had gone blind in one eye. The artist had simply opted for realism over Pharonic dignity. The prevalence of eye disease in ancient Egypt gives credence to this claim, as does the unique and innovative style of this artist, Thutmos. The graceful curve of the long neck, the arched eyebrow, the hint of a Mona Lisa smile on her full lips is a far cry from the symmetrical frozen immobility of the traditional Egyptian statuary of the time. This view too had to be abandoned, however, when new wall reliefs and other three dimensional figures were found. Some of these were clearly by the same hand that had carved the famous bust, that of Thutmos. These show the queen at a later and older age with both perfectly good eyes. No satisfactory consensus was ever reached to explain this intriguing archeological mystery.

THE FUTURE

The old Biblical Prophecy had come true and the meek had finally inherited the Earth, albeit a meek and empty one. There is no tooth nor claw nor danger here anymore and nor are there any lions left to lie with lambs. The mighty Nile that once roared in a yearly flood to feed millions is now only a pitiful ditch in which an arthritic rat could not drown, if there were any rats left to drown, that is. Even this trickle, like all other once great rivers of the world, is slowly vanishing, its moisture evaporating drop by invisible drop, never to rain upon land or sea again.

With the exception of a few Caretakers, humans too have gone to the stars and are scattered among the galaxies. Caretakers are the descendents of those who did not leave with the original Diaspora but chose to stay on the dying Earth. Now they live in Biospheres; protected and pampered like hothouse roses, rarely venturing out into the thin anemic atmosphere; and then always covered head to toe for protection from the deadly UV. Ozone layer, as Nardan knew from his 'Earth History', was the first to go.

For reasons Viceroy Nardan was not privy to, the Alien Raj had turned the planet into a museum cum zoo and the Darbar treats it as if it were a precious egg. It doesn't seem to matter that the egg had long since hatched. What is left is only a shell; or rather, a combination of many shells: shells of empty decaying cities, empty crumbling buildings, empty oceans and a lifeless land with poisoned soil. Even the humans who stayed behind are kind of shells themselves, living mildly parasitic existence, pretending that their hobbies of cultivating orchids or breeding some variety of gecko or goat were important to the Raj. The Darbar (and who can figure out the Byzantine workings of the Darbar!) encourages this delusion with generous funding. Not that he was complaining. Baronetcy came with the assignment, a title reserved in the diplomatic corps only for those administering whole galaxies, not to mention a salary that will leave him wealthy on his retirement.

Biosphere 77, named Memphis after a city that was already ancient when Nefertiti was born, was a good example of Darbar's indulgence. It was the largest Biosphere on the planet, built in the shape of a pyramid. Inside its 64 square miles, no expense was spared to create full size replicas of many other old cities. And despite himself, Nardan ZXX-2, Alien Raj's Viceroy to Earth, (Imperial Administrator), felt a strange awe standing among the grandiose stone pillars of the Luxor Hall. The pillars, he was assured by his host the Khadive of Memphis, were the exact replicas of the colonnades of ancient Temple of Karnack. As a life long diplomat, he has visited many of the Great Worlds of the Empire and yet the power of this second hand pomp and pageantry of a primitive kingdom from this dying planet's past was undeniable. Once again, he wondered if he was beginning to go 'native'.

"I didn't know there was so much interest in Nefertiti." He said as he looked down on the overflowing hall. The whole hall was full of dignitaries from the other biospheres. And, not surprisingly, reporters were everywhere. The proceedings were, after all, being simulcast on Multi-Net.

"They don't care about Nefertiti". Said Gaballah Ali Gaballah, Khedive of Memphis, "They have come to witness a public suicide, to see a man jump off the world." His voice was somber as a politician's at a funeral, matching the mood of the crowded hall.

The Viceroy understood the Khadive's dilemma. If there was any meaning to be mined for existence on the now dead Earth, it all lay in the past. Yet time travel was a taboo on Earth. But then, what wasn't a taboo on this backward looking planet? The only reason there wasn't a taboo against space travel was that it had saved human kind from certain extinction. It was the discovery (some even say a gift from the Raj) of the Star-Gate Time Machine technology from Jack Sarfatti's 21st Century generalization of Einstein's unchanging cosmological *constant* to a changing cosmological *field* at the 11th hour that had rescued humanity from the dying planet, which was thoroughly trashed by then, sucked dry of all its fabled milk and honey. Had it not been for Sarfatti's Star-Gate technology, humankind would have perished by its own blind success. But the issue was more complex concerning time travel to the past. In addition to the taboo that even the Diaspora humans respected, it was, at best, according to many, a one way trip because that other physics genius, Stephen Hawking, back in the 20th Century; in his "chronology protection conjecture", had said that radiation in endless time loops would get infinitely blue shifted and burn the time traveler to a crisp at the precise moment that the time machine formed out of the Star Gate. This theoretical "Battle of The Titans" between Sarfatti and Hawking had never been settled experimentally. On the other hand Hawking was a sly fellow and would waffle a bit sometimes seeming to take both sides joking continually about "flying saucers". Kip Thorne, who made the first breakthrough in Star Gate physics in 1986, had a failure of nerve on this issue of time travel to the past although he did bet Hawking it was possible, he never thought he would really win the bet. Igor Novikov, however, who independently of Sarfatti, also conceived of the "Destiny Matrix" or "globally self consistent" paradox-free time travel to the past, with "destiny" limiting "free will" always thought it would be achieved. These debates were, of course, before the Alien Raj made their presence on Earth public in 2012. There was also the constraint that one could not go back in time to a time before the Star Gate was formed. It was only because of the recent contact with a very ancient alien civilization that had old enough Star Gates that this present expedition to Ancient Egypt became thinkable. For some strange reason, the aliens would not discuss time travel to the past saying only "Seek and Ye Shall Find". The Viceroy, however, had no choice but allow Tom Sefari his right to go. The Alien Raj Constitution was unambiguous on the matter. "All citizens of sound mind are forever free to travel to the many parallel material "brane" worlds of Super Cosmos. Up until now, shorter time travel trips to the past had been attempted by humans but no one ever returned. Trips to distant parts of our universe and to the universes next door were, of course, common place, but in every such case, no humans, and none of their alien acquaintances with whom they traded goods and services, had ever returned to times that were earlier than when they left on the trip. Indeed, this would allow a younger person to meet his older self!

"Here he is"! Whispered the Khadive, pointing to a man walking towards them. There was no need to point him out; there was no mistaking him. He was a handsome human; surprisingly athletic looking with strong callused hands, dressed in the Egyptian fashion of sixty centuries ago. Nardan was told that the fabric was made from cotton grown from the original heirloom seeds, hoarded and protected for all these hard starving centuries. Once again he felt a certain awe at the impeccability of these people's attachment to their past. Perhaps it was a good thing that they had taboo against time travel.

"He seems in good physical condition to withstand the rigors of primitive society. I have read about the infectious diseases of your past." Said the Viceroy.

"O yes, he is quite fit; and more than just physically. Even lacking the spoken language, he should be able to fit right in. He has skills that are valued by that culture. He was my student, you know, taught him his hieroglyphics, so to speak. As to the diseases, we have inoculated him against most common ones of that age. I only hope that he will get the chance to need it. There is no proof that any of these trips have succeeded even one-way. Certainly, no one has returned."

And then, to the sound of funereal wailing of trumpets, Tomas Sefari walked into the Time-Gate to the foggy past to seek answers to an ancient mystery.

THE PAST

Tomas Sefari hovered in the clouds over a bluff, looking down as if he was god Horus the hawk. Below him the Nile was a wide ribbon of shimmering water speeding to meet the Mediterranean. A sea of wheat fields spread on both of side of the river as far as the eye could see; some harvested, tied in neat bundles and arranged in straight rows, others ripe and ready, yellow as the sands of Sinai. The sky was turning pink in the East; the sun will soon rise. His feet felt the ground and ripe wheat stocks blown with the gentle breeze rubbed his ankles. He was in the land of the Pharaohs. The 'River of Time' has brought him here to the land and time of his dreams, the time of Nefertiti.

He was here but he was not himself. He has ceased to be Tomas Sefari, once upon a future archeologist. He was now Thutmos the artisan. Perhaps it was always so, perhaps, he was always Thutmos. Certainly, now, standing on the fertile land of Sacred Egypt, he was Thutmos the artisan and no one else.

A boulder polished by ages protruded like a tax collector's heart in the middle of the wheat field. Here the plowman had connived with necessity and bent his furrows around the unyielding rock, creating a thing of beauty, a place of power, a place to welcome the morning sun, maybe even a time traveler. He walked towards the boulder as if pulled by it.

The boy bolted from under the rock; sleep still in his eyes, and fear. Thutmos grabbed him and held him, smiled at him and offered him a piece of bread, the universal symbol of peace. Where had the bread come from? Of course, he does not remember, nor does he remembered the young student named Javed Chaudari who had helped him prepare for the trip, the trip everyone else in Memphis thought was just a disguised form of suicide. Javed, while packing tools, through some strange clairvoyance, had also added a loaf of hard bread to the traveler's kit; and now that bread, baked in a far future oven according to an ancient Egyptian recipe, help break the language barrier for the two strangers.

Fate rules the world. Never were two people better met than Thutmos and Jabedi, a time traveler and a run away slave. Jabedi taught Thutmos the Theban language, guffawing at his mispronunciations; Thutmos taught Jabedi to carve soapstone into figures and to whittle faces from the driftwood. They wandered along the Nile, always going north, Thutmos prompted by some subliminal longings, Jabedi just to get farther and farther away from his sadistic owner. Almost preordained, their friendship deepened.

One hot afternoon, thirsty and tired from the dusty road, they came upon an inn. The walled courtyard with its gate wide open was inviting. Men snoozed in the shade of trees on thick papyrus mats or sat talking and drinking beer made by the innkeeper. Tethered donkeys dozed under palm trees, resting only as a beast of burden can. In the shade of an old fig tree, a woman cooked in an outdoor kitchen.

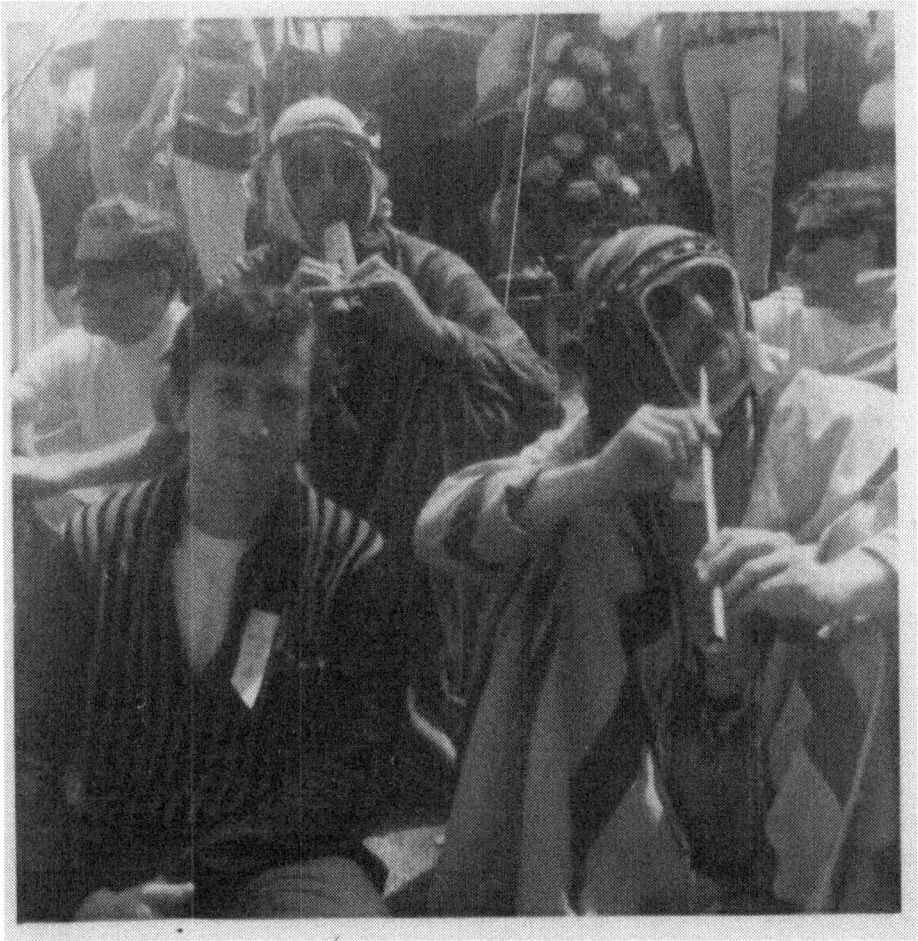

Thutmos Sefari at the Inn

Emboldened by his hunger and taunted by the aroma wafting from the flaming brazier, Jabedi walked up to her.

"Mistress, for the price of a meal, my master will draw the likeness of thy face which is more beautiful than the full moon of the harvest season". Said Jabedi, his voice a dry croak.

"Drink!" she pointed to the water pitcher. "Drink some water before you flatter me or sell your master; and give him some too".

Thutmos did not draw the likeness of Dulea, the cook and the proprietress of the inn; he painted it in bright living colors on the huge adobe wall of the inn overlooking the Nile.

In the mural, Dulea stands in her outdoor kitchen, the brazier flaming next to her. The fig tree, with its lush green leaves and pregnant with fruit, fills and forms the background. Bundles of onions and garlic hang from the wall. Foods of all sorts: a roasted duck, a fish on a hook, a basket of bread and bunches of fruit sit on the raised counter. Surrounded by this cornucopia, her arms full of green vegetables, she looks abundant, like an Earth Mother. And you could see the mural from miles. Business of the inn boomed. Dulea was the greatest fan of her own picture. She found many reasons, some quite odd and others totally transparent to be on the river in her little boat.

"No matter where you go, the eyes follow you". She said proudly.

"She is happier than a Pharaoh's daughter". Said her taciturn and adoring husband, Juma.

And then the Pharaoh's soldiers came. Thutmos was putting finishing touches on his second mural for the Inn, a bird's eye-view of the Nile, this one painted on the front wall. It shows Juma standing next to the waterwheel that hauls the river water for the use of the inn's gardens and guests. Below him, filling the horizon is the Nile. On its broad back float sluggish barges and swallow swift sailboats; and anchored far in the distance, a royal galleon, colorful as a firecracker.

With the clatter of hooves, dust, and barking of orders, the soldiers rode into the courtyard. Fear hung sweat thick in the Inn. People herded themselves near the kitchen, vainly seeking security in the company of others just as helpless. The soldiers brought nothing but death and slavery to the common man.

"Where is the one who painted that?" demanded the officer of the horsemen, pointing with his spear to the wall mural. Not an eye raised or a muscle moved. Thutmos put his reed brush down in the water bucket by his side and looked up at the horseman, his artist's eye admiring the deadly craftsmanship of the spear-maker.

"Keep thy eyes cast down, O artisan. Provoke them not". Whispered Dulea to him, trying to be discreet but her panicked hiss heard by all but Thutmos.

"Foolish woman, bringing attention to herself". Thinks her husband, sensibly still and silent.

"It is I", said Thutmos, walking towards the officer. Jabedi thinking it was a stupid thing to do, followed anyhow, close on his heels, but still gave lip service to prudence and kept his eyes mostly on the ground.

"Are you also the one who pained the picture that can be seen from the river?" Inquired the officer, already, his voice had lost its harshness.

"That is so". Replied Thutmos, pride of his work in his voice.

To everyone's amazement, the officer climbed down from his horse.

"Sir!" he said, bowing, "You are asked to present yourself for an interview with a person of importance. A boat will collect you tomorrow morning at the Inn's docking. This tunic is sent for your use". He gave Thomas a bundle.

Hardly anyone slept that night. The inn hummed with gossip. Who was this person of importance? Who was Thutmos for that matter? All attempts to talk to Jabedi or pump Dulea for information failed, leaving a field for speculation rich as a compost pile.

"He is a banished Prince of India. Now that the cruel regent is dead, he is to go back and rule his native land and free his sister and mother from the dungeon where they have languish for many years". Said the storyteller with an entirely unjustified confidence.

But a less fanciful explanation offered by the donkey-driver, delivered with visual aids, held more sway.

"Look at my donkey over there, just look at her', he pointed to the donkey. "Now, look at this'. He passed around a small papyrus sheet. Thutmos had done the handsome animal justice; the charcoal drawing had captured the vitality of the young beast.

"As anyone with eyes can see, this man Thutmos is a damn good painter of pictures. The tax collector has heard of his talent. He is going to decorate his tomb. Even now, if you take the trouble to look, there is a galleon anchored on the Nile to take him to the city of the dead".

The donkey driver did not know how uncannily close to the truth he had come. It was not the tax collector but Nefertiti, the queen of Upper and Lower Egypt, who had seen the mural on the wall. The sun was shinning on the mural as the queen's galleon passed by. The peasant woman looked alive, her eyes following the royal barge. Painted leaves seemed to move in the breeze and the queen thought she saw a bird dive for the painted figs, mistaking it for a real fruit. Even Hetusha, her chamberlain agreed that yes, the wench's eyes follow her majesty with admiration.

"Bring him to me Hetusha," she ordered. "The man that painted that mural, bring him to the barge. Make sure he is treated with respect. No harm must come to him. We need men like him to build the New Capital".

Akhetaten, the new capitol of Akhenaten and Nefertiti, was the first planned city in history. In 1359 BC, one hundred thousand workmen and artisans arrived at the virgin site, a barren desert (Egyptians never used arable land for building) on the western bank of the Nile, midway between Thebes and Memphis. In less than four years, they built a city with broad avenues, lush gardens; great temples dedicated the sun god RA. There were elegant homes, ostentatious palaces, all built to last. That they did not was not the fault of the builders.

Thutmos was the master sculptor of the 'city of dreams', and the most favored of its queen. And he loved her with all his heart. No woman was ever painted or sculpted with more passion for perfection than Nefertiti by Thutmos. He sketched her as she posed for him in her different and mercurial moods, he chiseled her likeness in alabaster, granite and sandstone. As he knew her more, his art transcended tradition and time and touched the face of eternal beauty.

The Face of Eternal Beauty

On the thirteenth day of Peret, in the year 1355 BC, Jabedi brought a load of sandstone blocks to the workshop. One small block (2'x2') took Thutmos's fancy. He carried it to his private workplace. His body shivered with the cool touch of the stone, his mind filled with strange premonitions. He closed his eyes and clearly saw the statute imbedded in the stone.

For the next two months, in secret, he coaxed the stone chisel stroke by gentle chisel stroke to release the face hidden within. Slowly the graceful neck, the high cheekbones and the curved brow began to emerge, and the sensual lips of the queen started to smile at the artist. Still in secret, he painted the bust with loving care. But the statuette stayed sightless. Thutmos could not bring himself to insert the eye inlays.

Finally, when he could find no more excuses, he forced himself. His hands shook as he put in the right eye retina; and as he picked up the piece for the left eye, his mind whirled, his head spun; and then, memories came flooding in, memories of another time, and memories of who he once was and why he was here. Thutmos/Tomas Sefari realized that he must not; he could not complete his masterpiece, his most famous work. He knew why he must not insert the left eye, why he must not disturb Destiny's Matrix and break the link of mystery between the Past and the Future. He walked to the river. The Nile was in flood and the little paste was never found.

Sleep Beauty for 3300 Years Till The Great Awakening

Joe Firmage and ISSO

Jack Sarfatti, Jagdish Mann, Creon Levit, Faustin Bray, Jean Pierre Vigier, 1999

Faustin Bray and Jack Sarfatti at Joe Firmage's "Last Hurrah"

Sound Photosynthesis

http://sound.photosynthesis.com/index.html

I forget who first put me in touch with Joe Firmage in I think it was 1998? Joe Firmage was about 30 and made a few million dollars in the Internet Dot.Com Bubble. He comes from a Mormon background descended from Brigham Young. Joseph Smith, founder of the Mormons, had an ET Contact like Abraham, like Moses, like Mohamed and others. So did Joe. Joe gave me money as a consultant to help him set up his International Space Sciences Organization (ISSO). I arranged for Creon Levit to join ISSO as "Scientific Director". Creon took a leave from NASA to do this. ISSO operated for about two years before going down in flames when Joe lost all his money. However, we did do some good scientific work on how flying saucers might work. We also debunked every "over-unity" bogus pseudo-science "free energy" claim that came to our attention. Creon set up a laboratory to test these claims. None of them worked. We had several theoretical physics conferences including a two-month visit by Louis De Broglie's[231] assistant Jean Pierre Vigier who was also a hero of the French Resistance in WWII and a close personal friend of Ho Chi Minh.[232] We also did a "classified" engineering project with SARA[233] Corporation in Huntington Beach, California and a table top hot fusion project with John Brandenberg on the East Coast in Washington DC and Princeton. The ISSO

---

[231] Predicted wave nature of electron using Einstein's special relativity and Planck's quantum formula.

[232] Vigier assisted Ho Chi Minh in combat operations in Vietnam which understandably angers Stephen Schwartz.

[233] Science Applications Research Associates work includes "nonlethal weapons Colonel John Alexander consultant." Former SARA Chief Scientist, James Corum, publishes on reality of "The Philadelphia Experiment". Corum is now Chief Scientist at Institute of Software Research in West Virginia funded via Senator Robert Byrd's direct efforts. Corum, who visited ISSO, is working on "electromagnetic stress" propellantless propulsion motivated by the flying saucer phenomenon. Hal Puthoff has refuted Corum's formula for electromagnetic stress propulsion, which in any case even if it worked, would not be real electro-gravitic metric engineering of a weightless warp drive without g-forces. The physical barrier from the weakness of classical of classical Newtonian gravity to achieving this seemingly impossible dream is the spacetime stiffness barrier of one Planck length per Planck energy = 10^{-33} centimeter warp distortion per 10^{19} Gev applied energy expended. My Feynman Z renormalization factor (see technical papers at end of this book) may solve that problem. Too soon to know for sure. James Woodward, a physics professor at Cal State agrees with Puthoff on Corum but like me, disagrees with Puthoff's "PV" theory of gravity and with Puthoff's attempt with Bernie Haisch to derive gravity and inertia from the transverse zero point photons.

Theoretical Physics Group included Saul-Paul Sirag and Dr. Vladimir Poponin from Moscow, Russia.[234]

---

[234] I do not think Joe's current efforts to keep the ISSO program going in "Motion Sciences" will succeed for two reasons. First, the physics theory of Haisch, Puthoff and Rueda, behind it is not at all sound in my opinion and has no chance of working as advertised. They only have a fragmentary view of the whole. It is too incomplete. Second, the current world economy is too depressed with the bursting of the Dot.Com fantasy and with the terrorist attack on the New York World Trade Center. Motion sciences has already gone kaput as has Bernie Haisch's Laputan CIPA funded by Joe to "make sun beams from cucumbers" with *uncontrollably random* zero point photon energy (as in the SED model) in the flaky sense of Nick Cook's silly book "The Hunt for the Zero Point". The NIDS UFO operation has been scaled back. Joe's latest attempt to recoup his lost fortune is OneMany.com. I am told that the software is actually quite good if it can really be made to work. There are serious problems with it and with the business plan. Bernie Haisch is the "Science Officer". What that means exactly I do not know. I was and am upset with Joe's stubborn Einstein-bashing and with Bernie Haisch's tolerance of Joe's incorrigible crackpot views on Einstein's physics. Hal Puthoff wrote me that he has been trying to educate Joe about physics these past two years since I gave up on that hard task. Judging by Joe's latest pronouncement in August 2002 that Einstein's general relativity lacks adequate experimental confirmation, Joe has still not passed the course. Part of this is Hal's fault since Joe is basically citing a paper by Dicke written in 1961. This paper is where Hal got his basic PV equations. has was remiss in not telling Joe that the paper is considered obsolete today especially in regard to the experimental tests of Einstein's physics in the past 40 years. Joe approaches science as if it were ideology and theology. This is dangerous for a man still young with his eyes, and his father's eyes, on the White House in say 20 years. Joe sees himself as a world leader meeting with Presidents and the Dalai Lama. Joe, as a descendent of Brigham Young, may be able to get Mormon money to back him. It is alleged that he also has the support of international wheeler-dealers like Maurice Strong. The problem is that Joe, in my opinion as an Ivy League Cornell graduate and in my working with him, has not been properly educated and does not really know what real knowledge and critical thinking are. He has naively and fanatically bought into all the mindless superficial est-Esalen-State of the World Forum toxic Cargo Cult pseudo-scientific psychobabble and sham-spiritual cosmologies of the California New Age Consciousness movement that emerged from the 60's New Left. This could be very dangerous for America and the world in the future. Conscience compels me to blow the whistle and sound the alarm. If Joe were to spend say four years at Yale or some such equivalent East Coast Ivory Tower, there might be redemption for him. Charisma and just enough intelligence to manipulate men and women without real education are very dangerous. I already saw this same thing up close and personal with Werner Erhard 25 years ago. This is my report to the Nation.

Vigorous Vigier at age 81, Viva La France!

Cold Fusion? ISSO Science's Position by Creon Levit

Creon Levit wrote[235]: "There are several questions here. The first is the question of the reality of low energy nuclear reactions (including cold fusion and other phenomena), the second is the viability of the claims, and finally there are the political issues. I have been interested in this for a while. I followed it fairly closely for a few years, but was never really "in the field" and have been paying less attention for the last year, mostly due to other demands on my time.[236] For what its worth, here is a brief summary of my positions on the questions.

---

[235] On December 12, 2001

[236] As day-to-day director of ISSO Science coordinating program with SARA and other projects.

On the reality of low energy nuclear reactions, I personally think there may well be something there. But the field is not healthy. This is because it is heretical, the initial announcement was problematic, the replications are troublesome, and what little objectivity there was or is usually swamped by polemic from true believers and debunkers alike. Also, the brave academics that embarked upon serious research in this area were usually already tenured, since only they did not have to worry how publishing heretical material would affect their careers. But this happened over ten years ago. Many of them are becoming less active because of age, and there is very little fresh blood being added from the main stream for obvious reasons. It is very hard to get funding for this stuff, unless you get it privately by making big economic promises.

It does seem to me that the initial replications by the big names (MIT, Princeton, Los Alamos) were not objective or careful enough, and Eugene Mallove makes a good case that at least for MIT, there may have been, shall we say, some "scientific misconduct". He was a skeptic who went in to investigate but eventually resigned from MIT because of what he found. I also agree with Mallove that the initial DOE report dismissing the phenomena is not fair.

But as yet, so far as I am aware, there is no example of an electrolytic or gas phase experiment that produces unambiguous nuclear reactions, be it heat, fusion products, or transmutation, and that has been replicated by more than one group. There are literally dozens of experimental claims, some from reputable institutions, that put forth positive evidence in one or more of these areas, but with the possible exception of some work at SRI, I know of no exact replications that came out positive.

As for the viability of Black Light Power's claims, I am not an expert. Some time ago, if I remember correctly, they backed away from their energy claims and were promoting their "exotic chemistry" claims instead. What about claims of sub-ground states of hydrogen (hydrinos) in chemical combination with metals? This in and of itself would still be the biggest news in physics for a long long time, if it were true. I'm not sure if it is or not but they are not the only ones

193

who believe such a thing is possible.[237] Are they promoting over unity energy again? I should have a look.

Hal[238] and Scott and company at Earth Tech have attempted some replications of the work of Black Light, as well as of Storms, of Mizuno, and of other big names in the cold fusion community. You can find their results on Earth Tech's web page. Earth Tech is highly competent, productive, funded for the "right reasons", dying for a positive result, and they have no economic conflict of interest regarding the outcomes of their experiments. They are, primarily, honest. But they have no breakthroughs, or even positives, to report at this time so far as I know. Mallove himself has admitted to me that what he wants is an experiment that convinces Hal Puthoff's lab. I don't think he has found one.

But I don't want to appear too negative. The cold fusion phenomenon is elusive, the funding is scarce, and it is not a good career move. That really is too bad. Work in this area is too important to let it be dictated by fashion or forgone conclusions. Personally, if I had the wealth of some of today's newsworthy philanthropists, I would spend a few million a year on cold fusion. If I were the NSF, I would do the same. There's nothing like an announcement of a well-funded peer-reviewed competitive grant program to get both nay Sayers and new folks to take a field seriously.

If I were on such a review committee, I would look favorably upon proposals that utilize ion beams bombarding hydrogen loaded metal foils in vacuum. This gets rid of a lot of the alchemy. There have been interesting Japanese results in this area. Also, I believe there is good evidence for unexpected transmutation in some of Ken Shoulders' charge cluster[239] experiments. Why no one is attempting to replicate any of these experiments is no mystery. There is no money available to do it. Tragic. [240]

---

[237] J.P. Vigier has such a theory and experimental work on it has been going on in Belgrade for years. Creon met with these Serbian physicists with Vigier in Budapest. Creon and Vladimir Poponin also went to Moscow to meet with several physicists. Creon also met with Hagen Kleinert in Berlin. This was a very important meeting because that is how I learned about Kleinert's idea of the "world crystal". Kleinert studied with Feynman at Cal Tech.

[238] Puthoff

[239] A billion to a trillion electrons seem to clump together in defiance of the Coulomb repulsion. Skeptics dismiss them as charged mercury droplets. The debate is unresolved. One possibility is that the $\square$ field is negative inside the hollow cavity. This would, perhaps, give a strong short range "Abdus Salam" gravitational attraction stabilizing the charge cluster. The same idea on a larger scale is "dark matter". Therefore, the charge clusters, if not mundane charged mercury droplets, might be small samples of "dark matter".

[240] Joe fumbled the ball on this one - this is Jack's opinion not Creon's.

If you want to get up to speed quickly, order the "10 year anniversary of cold fusion" issue of Mallove's magazine "Infinite Energy". It is a little dated, but well worth the couple of bucks he asks for it. Also, there are several conference proceedings; probably available through "Infinite Energy" though not published by them that contains some extremely interesting papers, as well as some junk. Though Mallove now seems firmly in the true believer's camp, he at least has a right to be.

As for President Bush not being a cold fusion proponent, after all the money he probably just lost on Enron, I would think even he's looking for an alternative to the oil business. My guess is he probably just does not know.[241] As they say, "Never ascribe to malice what simple ignorance suffices to explain".

---

[241] Jack adds: also he is too busy now with Usama bin Laden and al Qaeda to worry about cold fusion and flying saucers. Maybe the Grays will abduct bin Laden? ☺

***Jagdish, Vigier and Jack, Telegraph Hill, San Francisco, 1999
at ISSO Live-Work Space***

Saul-Paul, Jack, Tina, Vigier, Eldon Byrd [242]***, ISSO Telegraph Hill, 1999***

---

[242] Eldon Byrd is an engineer working for US Navy who investigated Uri Geller in the 70's along with Wilbur Franklin who had a strange death.

ISSO/ISEP[243] Research in the Physics of Consciousness

Dick Bierman and Jack Sarfatti at ISSO Telegraph Hill[244]

http://www.fourmilab.ch/rpkp/

Evidence for backwards in time "retro causality" mental effects in our brains.

Related links

http://noosphere.princeton.edu/

Global Consciousness Project[245]

---

[243] "Internet Science Education Project", http://stardrive.org/
[244] The white board shows the experimental pre-sponse effect in our living brain of future causes of past effects. See Dean Radin's "The Conscious Universe".
[245] "The analysis of data from September 11 has some new aspects. We have now established that the strength and persistence of the effects in the EGG data are unique in the three-year database. Focused analyses of correlations across the eggs, moment by moment, as well as correlations across time, as the day progressed, both show extraordinary, highly significant relationships where there should be none." - Roger Nelson

http://www.datadiwan.de/SciMedNet/library/articlesN73+/N74Nunn_time
.htm

Explaining Radin's and Bierman's pre-sponse brain data.

"On the evidence so far, pre-sponses are genuine causal anomalies that appear to be readily reproducible. If these characteristics turn out to be robust, they'll have to be explained. Major questions centre on whether orthodox quantum theory can be modified to accommodate them and, if so, what sort of modification is needed. Some quite intensive discussions on this topic were held on-line (in the 'quantum mind' forum) at the end of last year. Two front runners emerged from the discussions, one due to Henry Stapp and the other to Jack Sarfatti, both of whom are physicists based in California."

Henry Stapp's Bohr-Heisenberg idealistic model of pre-sponse

"Stapp's idea is to push the Copenhagen interpretation of quantum theory as far as it will go. This leads him to a view similar to that of Alfred North Whitehead who held that reality is a series of 'knowings' (Whitehead sometimes called them 'actual occasions of experience'). In particular, until an event has registered in someone's mind, its nature is not necessarily fixed. So the apparent pre-sponse could be an outcome of the unfixed nature of reality prior to observation by someone (note that the GSRs were recorded and were not seen by anyone until long after each experimental session)."

Jack Sarfatti's Bohmian realistic model of pre-sponse

"Sarfatti's suggestion is quite different. He builds on David Bohm's picture of the quantum world, which divides reality into 'particles' and the 'quantum potential'. According to Bohm the 'quantum potential' has no sources and thus cannot be influenced by the 'particles', despite being responsible for guiding their motions. Sarfatti however suggests that the 'particles' of certain systems, including brains, can feed back to influence the 'quantum potential' in a manner which essentially involves loops in state space (i.e. imaginary space) but can sometimes manifest as temporal loops in the perceived world." - quotes by Dr. Chris Nunn

A biologist reading the online preview of this book wrote to me asking why I believed in Jesus. I answered:

What do you mean? Do you mean my online book "Destiny Matrix" (soon to be in paper)?

That last chapter is by my 88-year-old father Hyman Sarfatti not me.

I have nothing against Jesus of course, but I am not religious in the old fashioned way like my mystical Dad.

I am more in the tradition of English Skepticism and French Rationalism, Roger Bacon, Thomas Hobbes,[246] John Locke, Voltaire et-al.

His views should not be confused with my own.

I wanted him to see his ideas in print before he passes to the other side.[247] Although he, and my Mom, are both mobile, hale and hearty with no major ailments.

I have a physical theory of consciousness that produces plausible numbers for humans like ~ 0.3 seconds for the objectively clocked resolving time of a subjective inner conscious moment based on 10^{18} qubits in the brain - the actual number of hydrophobically caged electrons inside microtubules (see Roger Penrose's "Shadows of the Mind" and Stu Hameroff's web page). The power dissipation needed to maintain our stream of consciousness is only $\sim 10^{-2}$ Watts. All of these numbers are tied to the Hubble parameter H that, in my theory, is related to a local cosmological $\Lambda(x)$ field that is the cosmological consciousness pump!

---

[246] "Leviathan as Ahab's "Whale" in Moby Dick as discarnate intelligence—"Star Maker" "Cosmic Coincidence Control" (John Lilly)?

[247] Like Kip Thorne wanting gravity wave data for John Archibald Wheeler now ~ 92

The Right Stuff

This section is mathematical requiring a mature physical intuition of an experienced theoretical physicist and can be skipped by the non-mathematical faint-hearted reader.[248] ☺

The local cosmological field pumping our stream of inner consciousness is

$$\Lambda(x)_{here-now} = \left(\frac{H(x)}{c}\right)^2 = \left(\frac{1}{L_p^*(x)}\right)^2 \left[1 - L_p^*(x)^3 \left|\Psi(x)\right|_{future|past}^2\right]$$

$L_p^*(x)$ is the locally variable quantum gravity Planck scale *amplified* from hyperspace.

$\Psi(x)_{future/past}$ is the spontaneous broken phase symmetry *local* order parameter of quantum vacuum[249] whose partial cohering of the random zero point fluctuations of the false vacuum creates classical curved spacetime from Josephson phase modulation and the local cosmological field $\Lambda(x)$ from local amplitude modulation. The two are coupled together of course.[250] $\Psi(x)_{future}$

---

[248] "Faint heart never won Fair Lady." Iolanthe, Gilbert & Sullivan

[249] As in P.W. Anderson's "More is different" and as in the New York Times article on the challenge to particle physics from solid-state theory.

[250] Fred Alan Wolf's book "Star Wave" was an early attempt at formulating this idea. Fred and I discussed this emerging idea as far back as the 1960's when we were both physics professors at San Diego State. Fred's recent work in this, partially funded by my ISEP:
"A Quantum Physical Model of the Timing of Conscious Experience." in Toward a Scientific Basis for Consciousness, Eds. S. R. Hameroff, David Chalmers, and A. C. Scott. Boston, MA: The MIT Press
"The Timing Of Conscious Experience: A Causality-Violating, Two-Valued, Transactional Interpretation Of Subjective Antedating And Spatial-Temporal Projection" Journal of Scientific Exploration, Vol. 12, No. 4. pp. 511-542. Winter, 1998.
"The Quantum Mechanics of Dreams and the Emergence of Self-Awareness." in Toward a Scientific Basis for Consciousness, Eds. S.R. Hameroff, A.W. Kaszniak, and A.C. Scott. Boston, MA: The MIT Press, 1996, pp. 451-67.

is Aharonov's *retarded* "destiny vector". $\Psi(x)_{past}$ is Aharonov's *retarded* "history vector". The complex conjugate mirror images through the real number axis are the advanced state vectors as shown in John Cramer's "transactional interpretation" of quantum theory with its "handshake" between future and past. Cramer's "handshake" is essentially the same idea as Professor O. de Beauregard's "Feynman zigzag" which replaces faster than light quantum action at a distance by retroactivity from future to past, i.e. "telos".

Spirits of Christmas Future and Christmas Past, eh? ☺

Both past and future qubit inputs generate our subjective inner experiences of the "present".

$$N^{*2}\hbar \sim \left(\frac{e^2}{\hbar c}\right)\left(\frac{mc^2}{H}\right)$$

$$N^{*2}\ 10^{-27} \sim 10^{-2}\ 10^{-27}\ 10^{21}\ 10^{17}$$

$$N^{*2} \sim 10^{-2}\ 10^{21}\ 10^{17} \sim 10^{36}$$

$N^* \sim 10^{18}$ in agreement with the observed anatomical number of caged electron SETs[251] in the human brain

This critical threshold for the formation of "pre-sponse"[252] is

Post-Quantum Backaction ~ Quantum Action

or

It takes two to tango.

"On the Quantum Physical Theory of Subjective Antedating." Journal Of Theoretical Biology, 1989, Vol. 136, pp. 13-19.

[251] Single Electron Transistors at the nanometer scale.

[252] Post-quantum signal nonlocality violating orthodox quantum theory because of non-equilibrium at the sub-quantum level in the sense of the papers of Antony Valentini. http://www.fourmilab.ch/rpkp/valentini.html_ http://xxx.lanl.gov/abs/quant-ph/0203049

or

When you look into the abyss, the abyss looks back at you.

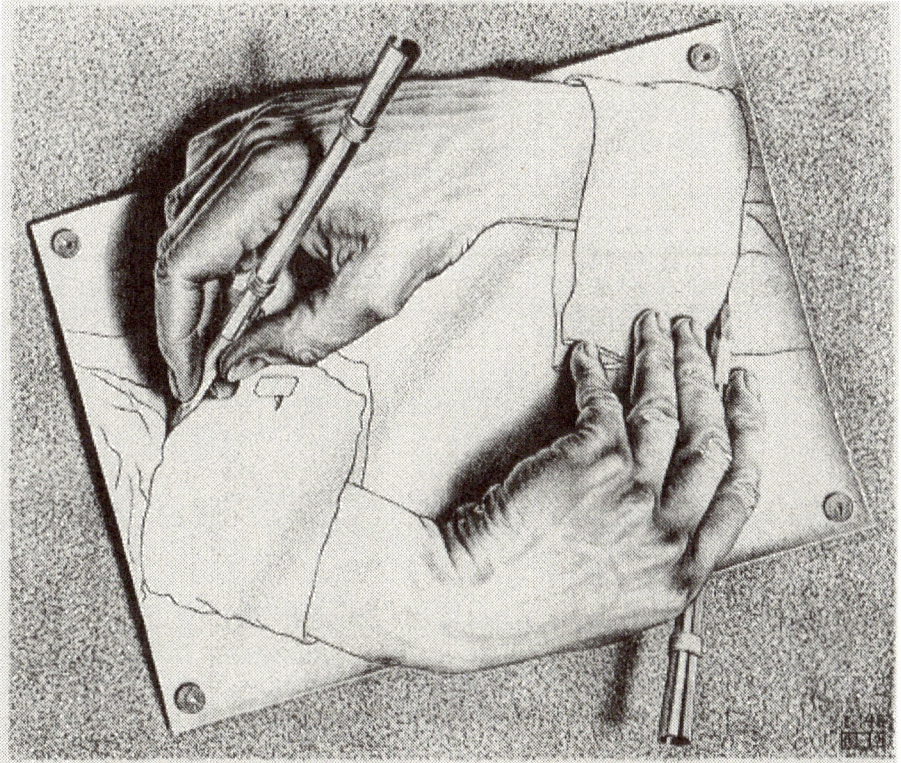

In orthodox quantum theory, N is so small that the backaction rate per quantum is too weak compared to the action rate, i.e.,

$$\alpha \frac{mc^2}{N^2 H} \gg \hbar$$

Post-Quantum Backaction Barrier Height >> Quantum Action Barrier Height

$$T \approx \frac{1}{HN*} \approx 0.3\,\text{sec}$$

In general this is

$$T \approx \left\langle \frac{1}{Nc\sqrt{\Lambda}} \right\rangle$$

For the duration T of moment of inner consciousness.[253]

$$N \sim \iiint dxdydz \left| \Psi\left(here-now\right) \right|^2_{future|past}$$

$$\frac{N}{\iiint dxdydz} \approx \left\langle \left| \Psi\left(here-now\right) \right|^2_{future|past} \right\rangle$$

$$\frac{NL_p^{*3}}{\iiint dxdydz} \approx \left\langle L_p^{*3} \left| \Psi\left(here-now\right) \right|^2_{future|past} \right\rangle \equiv L_p^{*3} \rho_s \left(here-now\right) \equiv \xi$$

Where ρ_s is the density of the superfluid in the partially coherent quantum "ground state" of the conscious matter. In the case of the quantum vacuum as the "ground state", the superfluid is "virtual", i.e., coherent quantum states of huge numbers of off mass shell virtual bosonic quanta. $\xi > 1$ is "anti-gravity" and $\xi < 1$ is "dark matter".

More details on this are below will be in my second book "Super Cosmos", which is the technical sequel to this popular "nonmathematical" book. The

---

[253] Chapter 14 of "The Undivided Universe" by Bohm and Hiley on the GRW model.

notation is explained more below. The symbol $\langle \, \rangle$ means take the appropriate average over the region where the qubits are roughly localized, e.g. the volume of the human brain with a billion billion (10^{18}) qubits. The "memes" are there all right. Their "qubit pilot wave" footprints in matter are in the synchronized spatio-temporal correlations of the material brain activity. These nonrandom correlations can be compared to a dance by a Corps de Ballet, which is how Bohm and Hiley picture superconductivity in "The Undivided Universe". These equations besides explaining consciousness, also explain, I think, how to make "vacuum propellers"[254] (http://www.fourmilab.ch/documents/vprop/) for flight without ejecting propellant (Bianchi identities on the quantum coherence corrected local Einstein geometrodynamic field equation), how to make time machine star gates connecting the universes next door to us in what Stephen Hawking calls "O Brane New World" ("The Universe in a Nutshell") and what dark matter really is.

It's a unified simple pretty picture unifying consciousness with gravity as Roger Penrose sees through the glass darkly and as we also did back in 1974 in "Space-Time and Beyond" - a silly immature work of course. We do not need the "Flat Earth Physics" of Haisch, Rueda and Puthoff (HRP) that corresponds to the "false vacuum" where $\Psi(x) = 0$.[255] In that case gravity in the form of classical curved spacetime cannot even come into being because the negative antigravity quantum pressure from the zero point fluctuations of the gauge force and source fields is too large and overpowers their positive zero point energy density fluctuations. You need to break the internal phase symmetry to compensate this effect. When you do that you get classical gravity in a globally self-consistent way like Andre Sakharov hoped for in 1967.

---

[254] Term coined by John Walker
[255] HRP assume a globally flat spacetime whose zero point dielectric response imitates gravity. They are missing the all-important coherence order parameter. The HRP theory is internally inconsistent. Also Puthoff makes no explicit connection of his classical "PV" model with quantum electrodynamics.

Telos as Final Causes in Post-Quantum Physics

"Like most of modern science, which characteristically shuns the deeper questions of 'final causes' and human purpose." Colonel Robert Hickson (January 6, 2001)

"Suppose there is even something vaguely teleological about the effects of consciousness, so that a future impression might affect a past action." Roger Penrose, "The Emperor's New Mind" pp 442-445 (1989)

"It seems to me that biological systems are able in some way to utilize the opposite time-sense in which radiation propagates from future to past. Bizarre as this may appear, they must somehow be working backwards in time." Sir Fred Hoyle, "The Intelligent Universe", p. 213 (1986)

No, I did not read Colonel Robert Hickson's paper of January 6, 2001 below before I wrote "Destiny Matrix" with allusions to Shakespeare's "Tempest" and Melville's "Moby Dick" and "Ishmael". I saw the work below for first time only on December 26, 2001.[256] This is John Lilly's "Cosmic Coincidence Control" in action, that "invisible Police officer of the Fates" in "Loomings" the first chapter of "Moby Dick."[257]

http://www.usafa.af.mil/jscope/JSCOPE01/Hickson01.html

"Strategic Defense-In-Depth of the Homeland and the Moral Hazards of Two New Materialist Ideologies of Sociobiology-and-Neuroscience's View of Man in Full and Their Likely Effects on the Moral Character of the Citizen by Colonel Dr. Robert Hickson, Visiting Professor, Joint Special Operations University , U. S. Special Operations Command"

---

[256] This reminds me of an incident from 20 years ago with I.J. Good who worked with Alan Turing cracking the Nazi war code. See Thomas Pynchon's "Gravity's Rainbow". Good wrote me a letter asking if I had read his "GOD(D)" paper that he gave in Chicago ~ 1980 to some parapsychology group. Of course, I had not seen it. The reason I.J. thought I had was that some of the ideas in his paper were identical to those I had in a paper in Psychoenergetic Systems. I.J. was miffed that I did not cite his paper, which I had not seen. These strong nonlocal correlations between my paper and I.J's had to do with a "superluminal telepathic discarnate intelligence" that Good called "GOD(D)" - very much like Stapledon's "Star Maker", Hoyle's "Black Cloud" and Melville's "police officer of the Fates" in Moby Dick.

[257] Thanks to Ron Pandolfi Ph.D scientist with the Central Intelligence Agency for alerting me to these considerations. Quoted with permission from Colonel Hickson.

Excerpts relevant to the thesis of this book:

"The neuroscientific view of life, has become the strategic high ground in the academic world, and the battle for it has already spread well beyond the scientific disciplines and, for that matter, out into the general public...

The new generation of neuroscientists...express an uncompromising determinism...[i.e.,] that there is not even any one place in the brain where consciousness or self-consciousness (Cogito ergo sum) is located. This [consciousness or self-consciousness] is merely an illusion created by a medley of neurological systems acting in concert...Since consciousness and thought are entirely physical products of your brain and nervous system—and since your brain is fully imprinted at birth [it is "an exposed negative waiting to be slipped into developer fluid" to develop what is "already imprinted on the film...i.e., the individual's genetic history"]—what makes you think you have free will? Where is it going to come from? What "ghost" [i.e., Geist, spirit], what "mind", what "self", what "soul", what anything that will not be immediately grabbed by those scornful quotation marks is going to bubble up your brain stem and give it to you? (Tom Wolfe, "Hooking Up"—Chapter 5, "Sorry, but Your Soul Just Died") [Given the inner logic of these neuroscientific premises, not only "free will", but also "free reason" itself, is an illusion—and a self-refuting proposition, to boot! -

Our provident and strategic (prudent and far-sighted) defense of the common good will be, however, an even greater test for us as a nation, in light of some very powerful, perhaps subtly ungovernable, and even dangerously self-replicating, new technologies, such as genetic engineering, bio-remediation, robotics, molecular electronics and other nano-scale technologies. Yet a far more important challenge to our "providing for the common defense" will be the intimate moral effects of certain shocking new views, which derive from our predominant scientific culture, about "the nature of man" and the world. Even within our predominant culture of scientific materialism, some of these new views are altogether shocking and psychologically dislocating—and politically "radioactive". Just as the theories of Darwin, Marx, Freud, and Nietzsche have had great effects upon man's perception of himself, and also upon the character of his pervasive political culture and even his strategic revolutionary warfare, so, too, will the growing new theories of sociobiology, evolutionary psychology, and neuroscience touch him most inwardly; and especially since they may also make intimately manipulative use of psychotropic drugs and GNR technologies, and other bio-technologies of power. Although, like most of modern science, which characteristically shuns the deeper questions of 'final causes' and human purpose—e.g., "What is man for?"—neuroscience and sociobiology will certainly compel us to re-examine the question "What is man?" Drawing on the language of the vivid-souled Tom Wolfe [2], we may frame our initial strategical admonition and challenge something like this: The "Confluence" or "Hooking

Up" of the sociobiological-neuroscientific view of life (and the natural man) with the self-replicating application of GNR technologies is a terrible thing to think upon![3]   The allure of hermeticism—the temptation of this potent hidden knowledge—and of unbounded demiurgic engineering, in combination, will be very great and intimately consequential.   Indeed, "If it weren't attractive, it wouldn't be a temptation."

And, it is hard to hold "the Natural Man" down.   The lure of unbounded liberation is seductive.   We may remember that Shakespeare's Caliban showed us that, very movingly.   Nature itself is not enough.   Neither is raw force, nor "power without grace", some would say!   Despite his beautiful poetry of heart and speech, Caliban himself was full of lust and spiritual astigmatisms.   He had, as it were, a restive libido dominandi; a desire for "power without grace"—and for the raptures that came from a delusive sense of liberation and dominion.   Let us think of utopian Caliban now amongst the scientists, and among the seductive new technologists, and we may see some of our strategic challenges more clearly...

In a famous essay by the British military historian, Sir Michael Howard, entitled, "The Bewildered American Raj: Reflections on a Democracy's Foreign Policy", [5] he notes our confused American dominion and "mood of resentment" (55) over the ingratitude of the world towards our power...

With these various considerations in mind from the political wisdom of our cautious Founding Fathers and from our disciplining Constitutional Order— while likewise mindful of America's doubtful cultural dominion, "Caliban's Bewildered Raj", and modern science's intrinsic and tendentious exclusion of telos and all Aristotelian "final causes"—a more descriptive and neutral alternative title for this essay might be: "The Great Power of GNR Technologies in Light of the World-View of Modern Neuroscience and Sociobiology, and their Combined Effect on the Moral Character of the Citizen and Honorable Fighting Man: A Strategic Defense-in-Depth of the Homeland under an Increasingly Centralized State and Balkanized-Uneducated Culture with Porous Borders".

But, first, another pre-emptive caution, and suggestive literary allusion.

A Message to my Gentle Reader:

If, unlike exile Ishmael on the first page of Moby Dick, you are unable first to "sail about a little and see the watery part of the world", thereby "driving off the spleen and regulating [your sluggish] circulation", you should certainly not too rashly read what follows!   For, the subject matter is (in the words of Master Rabelais) "a terrible thing to think upon"—especially if it is also now "a damp and drizzly November in [your] soul" and if you are likewise psycho-cramped and "grim about the mouth", as Ishmael alas was!   Rather, you should, like him, "account it high time to get to sea as soon as [you] can". And, I wish you good sailing! For the rest of you, we may continue."

End of excerpts from Colonel Hickson's paper.

Also quoted with permission from Neal Pollock; pollockn@spawar.navy.mil Chief Acquisition Engineer, PEO(IT)C1 http://www.usafa.af.mil/jscope/JSCOPE 00/Pollock00.html

"B.   People like to believe they are in control (i.e. conscious)

    1.   simple observation belies this belief
    2.   belief differs from knowledge; few study epistemology
    3.   people ascribe expertise to college degrees and job titles
        a.   most scientists have never studied the Philosophy of Science
        b.   understanding a specialty does not imply understanding per se
        c.   Our society supports a belief in causation—a bottoms-up approach—past drives the present

    1.   Jung developed synchronicity—meaningful coincidence
    2.   Jung spoke of a top-down approach, a teleological approach
        a.   the desired goal, for instance, drives the present from the future
    3.   when planning a journey you need both the start point and the end point
        a.   as the Mad Hatter told Alice, if you don't know where you're going, any road will take you there."

OK now to the physics explaining all of the above and refuting the "ant colony" view of Man by E.O. Wilson in his "sociobiology" and its derivatives all based on physics, bad and bogus. ;-) The key paper is by Richard Feynman:

"Space-time approach to non-relativistic quantum mechanics", Reviews of Modern Physics, Vol. 20, p. 267 (1948), Feynman [258]

---

[258] In Julian Schwinger's Dover volume on "Quantum Electrodynamics". p. 321 Equations (13) - (16) See also Feynman's Ph.D. on the forced quantum harmonic oscillator and the elimination of EM field oscillators to get both advanced and retarded light cone actions at a distance between electrons, i.e. Paper 23 in Schwinger.

"The quantity ψ defends only on the region R' previous to t...It does not depend, in any way, upon what is done to the system after time t. This latter in formation is contained in χ. Thus with ψ and χ we have separated the past history from the future experiences of the system...the asterisk on χ^* denotes complex conjugate...the function $\chi^*(xyzt)$ characterizes the experience...to which the system is to be subjected."

Feynman's Lagrangian method is precisely "a journey you need both the start point and the end point".[259] The key idea I present below is inherent in Feynman's Section 5 "Definition of the Wave Function" and it is the basis for Yakir Aharonov's "Destiny" and "History" two state vector "multiple time" version of quantum physics in which teleology (AKA "telos") is implicit. Teleology is explicated when I step beyond Aharonov from quantum to the post-quantum covering theory. This is a violation of "sub-quantum equilibrium" in the sense of Tony Valentini's papers.

http://www.edge.org/3rd_culture/bios/valentini.html

http://www.fourmilab.ch/rpkp/valentini.html

http://xxx.lanl.gov/abs/quant-ph/0203049

The post-quantum level has "signal nonlocality" in contrast to the quantum level, which has "signal locality"[260] in spite of "quantum nonlocality". Abner Shimony calls the latter "passion at a distance". It is a fainthearted kind of passion methinks. If Abner Shimony "is content with a vegetable love", which most certainly does not suit me, why he is welcome to it. ☺[261]   OK back to Feynman:

"Choose a particular time t and divide the region R into pieces future and past relative to t."

---

[259] I will discuss this in detail in my second more technical book "Super Cosmos".

[260] Quantum entanglement cannot be used as a stand-alone communication channel in orthodox quantum theory. It can in post-quantum theory. Pre-sponse is a violation of quantum theory. Pre-sponse is evidence for the reality of post-quantum signal nonlocality from local order parameters.

[261] http://members.aol.com/guron/pat/

The Cat's Cradle Principle: Collision of History with Destiny and Ice-9!

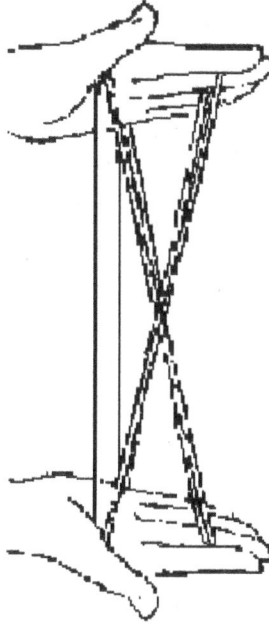

http://personal.riverusers.com/~busybee/catcradle.htm

The upper hand is the final cause hyper surface of future destiny. The lower hand is the past cause of history. I will now give my version of Feynman's insight using Dirac notation in a slightly novel fashion.

$$|\psi\rangle = |past\rangle\langle past|\psi\rangle + |future\rangle\langle future|\psi\rangle$$
$$= |history\rangle\langle history|\psi\rangle + |destiny\rangle\langle destiny|\psi\rangle$$

That is, imagine two spacelike surfaces "past" and "future" with all possible world lines that start on either surface and end up at "here-now" = spacetime event x. I generalize Galilean t to Einstein's 4-dim event x. For this discussion:

- "past" = "history"
- "future" = "destiny" or "teleology" or "purpose"
- "x" = "here-now"

Roughly as a first approximation. Therefore,

$$\psi(x) \equiv \langle here-now|\psi\rangle = \langle here-now|history\rangle\langle history|\psi\rangle + \langle here-now|destiny\rangle\langle destiny|\psi\rangle$$

$$|\Psi(x)|^2_{future|past} \equiv |\Psi(here-now)|^2_{future|past} \equiv \langle\psi|here-now\rangle\langle here-now|\psi\rangle$$
$$= \left[\langle\psi|destiny\rangle\langle destiny|here-now\rangle + \langle\psi|history\rangle\langle history|here-now\rangle\right]$$
$$\times \left[\langle here-now|history\rangle\langle history|\psi\rangle + \langle here-now|destiny\rangle\langle destiny|\psi\rangle\right]$$
$$= \langle\psi|destiny\rangle\langle destiny|here-now\rangle\langle here-now|destiny\rangle\langle destiny|\psi\rangle$$
$$+ \langle\psi|history\rangle\langle history|here-now\rangle\langle here-now|history\rangle\langle history|\psi\rangle$$
$$+ \langle\psi|destiny\rangle\langle destiny|here-now\rangle\langle here-now|history\rangle\langle history|\psi\rangle$$
$$+ \langle\psi|history\rangle\langle history|here-now\rangle\langle here-now|destiny\rangle\langle destiny|\psi\rangle$$

When one pictures this in terms of Feynman's "spacetime paths" or "world lines", it's a bit like a "Cat's Cradle" made from a string. [262] The nexus (crossing point) of the world lines is "here-now". [263] http://www.nytimes.com/books/97/09/28/lifetimes/vonnegut-cat.html

---

[262] Mentioned in email to me from Eugenia Macer-Story.
[263] Kurt Vonnegut's novel "Cat's Cradle" is a fictional account of my Cornell physics teachers in the Manhattan project. In addition to Colonel Robert Hickson's concern about a nano-goo destroying the biosphere, the quantum vacuum phase transitions are potentially equivalent to "Ice 9" if misused.

The Cat's Cradle is one of the meanings hidden in

Adam's hand is the history state vector pointing to the future. God's hand, the Finger of Fate, La Forza del Destino, is the destiny state vector, Henry Dwight Sedgwick's "home of explanation" pointing to the past.[264]   Do not confuse "past" and "future" with Wheeler-Feynman's "advanced" and "retarded". The Dirac "ket" $|\ \rangle$ is "retarded" and the Dirac "bra" $\langle\ |$ is "advanced". Therefore, the history and destiny state vectors each have advanced and retarded components. John Cramer's "transactional interpretation" is simply the limit of the model here when

$$\langle future|\psi\rangle \equiv \langle destiny|\psi\rangle \to 0$$

Therefore, in this generally wrong limit of retarded determinism of "sociobiology" and "strong AI":

---

[264] The opposite pointing hands also symbolize the post-quantum "two way relation" between pilot qubit wave and particle not found in the approximation that is orthodox quantum physics with signal locality. The "two way relation" (coined by Bohm and Hiley in The Undivided Universe) is the loop in state space mentioned by Dr. Chris Nunn above in addition to the time loops has signal nonlocality. The difference between quantum signal locality and post-quantum signal nonlocality violating the "no-clone" theorem is a key idea of this book.

$$\psi(x) \equiv \langle x | \psi \rangle \rightarrow \langle here-now | history \rangle$$

Is now a giant quantum vacuum wave function or local order parameter partially cohering locally random zero point fluctuations of virtual bosonic quanta of the "false vacuum" of the "unified field" in Einstein's sense and in the sense of what Andre Sakharov was seeking in 1967. I get $|\Psi(x)|^2_{future|past}$, in my theory of the local cosmological $\Lambda(x)$ field generalizing Giovanni Modanese's

http://int.phys.washington.edu/ect/postdoc/modanese/

http://xxx.lanl.gov/abs/gr-qc/9612022

with *four* kinds of globally self-consistent time loops. The two *off-diagonal* time loops in which history and destiny coherently interfere are obviously the most interesting. The naive retarded determinism of E.O. Wilson and his disciples is only one term out of the four. Furthermore, in the post-quantum domain the naïve determinism of "two new materialist ideologies of sociobiology and neuroscience" as well as "strong AI" in the sense of Roger Penrose, are replaced by spontaneously sentient self-organization in a Cat's Cradle of globally self-consistent Novikov "loops in time" of qubit flows symbolized of course by Escher's "Drawing Hands".

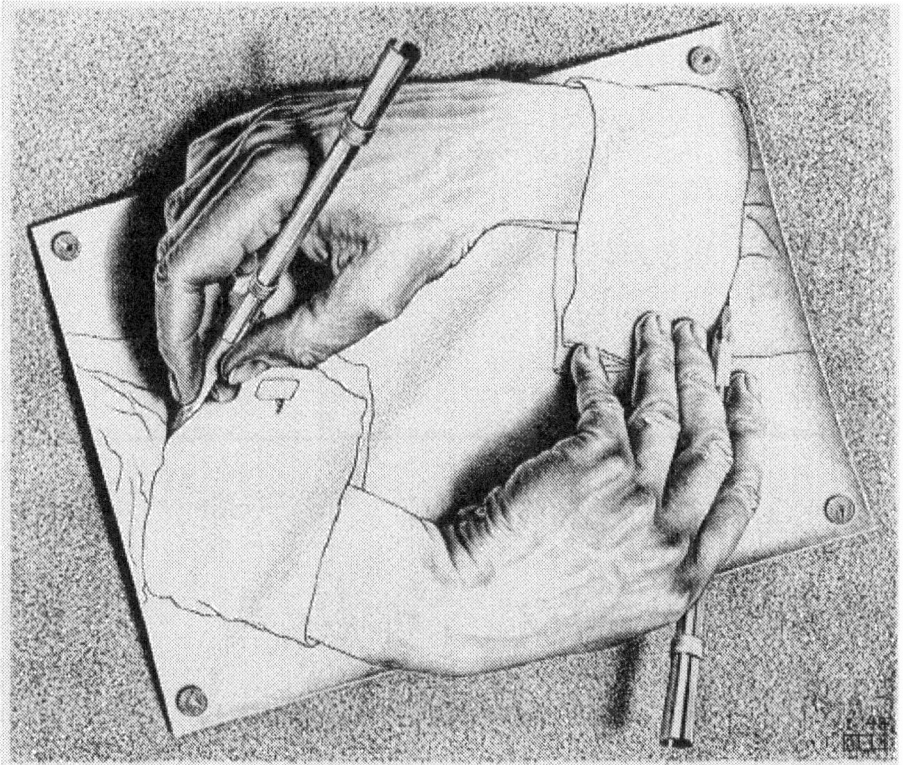

Spontaneously Sentient Self-Organization by the Time Loops of History's Collision with Destiny M.C. Escher's "Drawing Hands" © 2002 Cordon Art B.V. – Baarn – Holland. All rights reserved. www.mcescher.com

The quantum vacuum stress-energy density of this "post-quantum conscious computer" - we can call them "Spectra" class http://stardrive.org/cartoon/spectra.html ☺ is

$$t_{\mu\nu}\left(here-now\right)\big|_{virtual-quanta} = \frac{\hbar c}{L_p^{*2}\left(here-now\right)}\Lambda\left(here-now\right)g_{\mu\nu}\left(here-now\right)$$

Where $g_{\mu\nu}$ is Einstein's local classical geometrodynamical field.

$\Lambda(x) > 0$ is exotic quintessential repulsive "anti-gravity" for "vacuum propellers" and for keeping star gates stable and open.

$\Lambda(x) < 0$ is exotic attractive "dark matter".

The universe normally sits at $\Lambda(x) = 0$ which is Einstein's classical 1915 geometrodynamics.

Einstein's geometrodynamic local field equation generalizes to[265]

$$R_{\mu\nu}(x) - \frac{1}{2}\left[R(x) - \frac{2}{L_p^2}\left(1 - L_p^3\left|\Psi(x)\right|^2_{future|past}\right)\right]g_{\mu\nu}(x) = -\frac{L_p^2}{\hbar c}T_{\mu\nu}(x)$$

Lp*^{2} = area of one Bekenstein "bit" amplified by the universe's unseen dimensions of "O Brane World" (AKA "Super Cosmos") as in Stephen Hawking's "The Universe in a Nutshell".

---

[265] x = here-now

Send in the clones! The Achilles Heel of quantum cryptographic computing?

As I corrected this part about clones I heard a bell signaling an incoming
e-mail that was an advert for Star Wars Episode II 'Attack of the Clones'. Synchronicity strikes again!

http://www.vex.net/~buff/sinatra/cgi/arch.cgi/Send_In_the_Clowns [266]

The no-cloning theorem, not of humans, but of qubits, is the corner stone of the vaporware field of quantum computers, quantum cryptography and quantum teleportation.[267] The no-cloning theorem is thought to prevent the use of quantum nonlocality entanglement for stand-alone command-control-communication (C^3) both faster than light and backwards in time from future to past as actually seen in human brain pre-sponse retro-PK by Dean Radin and Dick Bierman. This is, therefore, a contradiction between theory and experiment. Strictly speaking the no-cloning theorem is formally correct. The problem is that it does not ask the relevant physical question. Let's review the conventional wisdom.

"If cloning were possible, then it would be possible to signal faster than light using quantum effects...to clone an unknown quantum state...is not possible in general in quantum mechanics..." A breakdown in the strict linearity of quantum superposition allows "time-travel", faster than light communication and violations of the second law of thermodynamics." The standard proof of the no-cloning theorem is irrelevant for a very similar reason to the irrelevancy of von Neumann's "proof" that orthodox quantum mechanics would not allow context-dependent nonlocal hidden variables.

Here is the standard proof that one cannot clone an arbitrary unknown quantum state. We want to "copy" or "clone" the qubit $|q\rangle$. The input state is

---

[266] "Isn't it rich, isn't it queer
Losing my timing this late in my career
But where are the clowns - send in the clowns", Stephen Sondheim
[267] How this new development on the intrinsic post-quantum nature of quantum computers of very high complexity affects the ability to quickly factorize large integers into primes to break RSA public key cryptography has not yet been investigated.

assumed to be $|q\rangle|s\rangle$, where $|s\rangle$ is the "target slot".[268] The quantum-computing gate is the linear operator U, i.e. unitary hence conserves probability, reversible etc. The goal is

$$U|q\rangle|s\rangle \rightarrow |q\rangle|q\rangle$$

Where $|q\rangle$ is arbitrary. Therefore, the dual relation on any other qubit $|q'\rangle$ is

$$\langle q'|\langle s|\tilde{U}^* = \langle q'|\langle q'|$$

Where $\tilde{U}^*$ is the complex conjugate of the transpose of rows and columns of the matrix representing the quantum computing gate in some computational basis i.e. quantum frame of reference in physically real, but immaterial intrinsically mental, qubit space beyond spacetime and matter. However, the quantum computing gate unitary matrix transformation U leaves the Dirac "bra-ket" inner product invariant because

$$\tilde{U}^* = U^{-1}$$

Where U^{-1} is the reversible inverse matrix that quantum erases the action of the quantum computer gate U. Therefore,

$$\langle q'|q\rangle = \langle q'|q\rangle^2$$

Hence, either

$$\langle q'|q\rangle = 1$$

Or

$$\langle q'|q\rangle = 0$$

Contradicting the premise that the two qubits q and q' are arbitrary relative to each other.[269] This is mathematically correct, but it is physically irrelevant, i.e. it does not ask the right question. It's the right answer to the wrong question.

---

[268] "Quantum Computation and Quantum Information" Nielson & Chuang, Cambridge 2000, p. 2-3, 18 and p. 532 Box 12.1
[269] Reducto ad absurdum

"What if we allow cloning devices that are not unitary? What if we are willing to allow imperfect copies that nevertheless are 'good' according to some interesting measure of fidelity? ...The cloning of non-orthogonal pure states remains impossible unless one is willing to tolerate a finite loss of fidelity in the copied state." p. 532 [270]

Knowing about "More is different" quantum optics, superconductivity, spontaneous broken symmetry in the quantum vacuum, (i.e., Goldstone theorem) leads to the following question. Phase coherent quantum states, like in atomic Bose-Einstein condensates, and in electrical superconductors, and in superfluid helium and laser light, are uncertain in the number of copies! We do not want eigenstates of the total number of copies operator in the Fock space with a fixed definite number of clones all marching in step, receiving their orders from a common pool of shared active quantum information! Indeed, the latter is mathematically impossible. For example, as a simple counter-example, we start with two target slots instead of only one. The *phase incoherent* input state is then $|q\rangle|s\rangle|s'\rangle$. What we want then is the partially phase coherent quasi "superfluid" output state

$$U|q\rangle|s\rangle|s'\rangle \rightarrow |q\rangle|q\rangle|s'\rangle\langle qqs'|U|qss'\rangle + |q\rangle|q\rangle|q\rangle\langle qqq|U|qss'\rangle$$

$$\left|\langle qqs'|U|qss'\rangle\right|^2 + \left|\langle qqq|U|qss'\rangle\right|^2 = 1$$

Similarly,

$$\langle q'|\langle s|\langle s'|\tilde{U}^* \rightarrow \langle q'ss'|\tilde{U}^*|q'q's'\rangle\langle q'|\langle q'|\langle s'| + \langle q'ss'|\tilde{U}^*|q'q'q'\rangle\langle q'|\langle q'|\langle q'|$$

Again the unitary inner product is invariant, however, this time, the spontaneous broken symmetry allows a non-trivial self-consistent finite polynomial (over the complex field) solution. Define the complex number z as

---

[270] This is an important potential loophole in how the Pundits interpret the no-clone theorem.

$$z \equiv \langle q' | q \rangle$$

The polynomial to be solved, with complex coefficients from the matrix elements of the quantum computing gate matrix U is then of the simple form

$$z = cz^2 + c'z^3$$

Let's prove this. Unitarity entails

$$\tilde{U} * U = 1$$

$$z = \langle q' | \langle s | \langle s' | \tilde{U} * U | q \rangle | s \rangle | s' \rangle = \langle q' | \langle s | \langle s' \| q \rangle | s \rangle | s' \rangle = \langle q' | q \rangle$$

$$z = \left[\langle q' s s' | \tilde{U} * | q' q' s' \rangle \langle q' | \langle q' | \langle s' | + \langle q' s s' | \tilde{U} * | q' q' q' \rangle \langle q' | \langle q' | \langle q' | \right] \times$$

$$\left[| q \rangle | q \rangle | s' \rangle \langle q q s' | U | q s s' \rangle + | q \rangle | q \rangle | q \rangle \langle q q q | U | q s s' \rangle \right]$$

$$= \langle q' s s' | \tilde{U} * | q' q' s' \rangle \langle q' | \langle q' | \langle s' \| q \rangle | q \rangle | s' \rangle \langle q q s' | U | q s s' \rangle$$

$$+ \langle q' s s' | \tilde{U} * | q' q' s' \rangle \langle q' | \langle q' | \langle s' \| q \rangle | q \rangle | q \rangle \langle q q q | U | q s s' \rangle$$

$$+ \langle q' s s' | \tilde{U} * | q' q' q' \rangle \langle q' | \langle q' | \langle q' \| q \rangle | q \rangle | s' \rangle \langle q q s' | U | q s s' \rangle$$

$$+ \langle q' s s' | \tilde{U} * | q' q' q' \rangle \langle q' | \langle q' | \langle q' \| q \rangle | q \rangle | q \rangle \langle q q q | U | q s s' \rangle$$

$$= \langle q' s s' | \tilde{U} * | q' q' s' \rangle \langle q q s' | U | q s s' \rangle \langle q' | q \rangle^2$$

$$+ \langle q' s s' | \tilde{U} * | q' q' s' \rangle \langle q q q | U | q s s' \rangle \langle s' | q \rangle \langle q' | q \rangle^2$$

$$+ \langle q' s s' | \tilde{U} * | q' q' q' \rangle \langle q q s' | U | q s s' \rangle \langle q' | s' \rangle \langle q' | q \rangle^2$$

$$+ \langle q' s s' | \tilde{U} * | q' q' q' \rangle \langle q q q | U | q s s' \rangle \langle q' | q \rangle^3$$

$$z = \langle q'ss' | \tilde{U}* | q'q's' \rangle \langle qqs' | U | qss' \rangle z^2$$
$$+ \langle q'ss' | \tilde{U}* | q'q's' \rangle \langle qqq | U | qss' \rangle \langle s' | q \rangle z^2$$
$$+ \langle q'ss' | \tilde{U}* | q'q'q' \rangle \langle qqs' | U | qss' \rangle \langle q' | s' \rangle z^2$$
$$+ \langle q'ss' | \tilde{U}* | q'q'q' \rangle \langle qqq | U | qss' \rangle z^3$$

$$c = \langle q'ss' | \tilde{U}* | q'q's' \rangle \langle qqs' | U | qss' \rangle$$
$$+ \langle q'ss' | \tilde{U}* | q'q's' \rangle \langle qqq | U | qss' \rangle \langle s' | q \rangle$$
$$+ \langle q'ss' | \tilde{U}* | q'q'q' \rangle \langle qqs' | U | qss' \rangle \langle q' | s' \rangle$$
$$c' = \langle q'ss' | \tilde{U}* | q'q'q' \rangle \langle qqq | U | qss' \rangle$$

These coefficients depend on the matrix elements of the quantum computer gate U and on the "initial conditions" $\langle s' | q \rangle$ and $\langle q' | s' \rangle$. It is quite obvious that any choice of U and any choice of $|q\rangle, |q'\rangle, |s'\rangle$, even in this simple low complexity case, always have non-trivial solutions for the invariant inner product constraint. I mean solutions that are different from $|q\rangle = |q'\rangle$ or $\langle q | q' \rangle = 0$. There is always a zero root of course. But in this simple model

$$c'z^3 + cz^2 - z = 0$$

$$\left(c'z^2 + cz - 1 \right) z = 0$$

Apart from the zero root,

$$z_\pm = \frac{-c \pm \sqrt{c^2 + 4c'}}{2c'}$$

"I'm very well acquainted, too, with matters mathematical,
I understand equations, both the simple and quadratical,

With many cheerful facts about the square of the hypotenuse."
Major General Stanley, Pirates of Penzance, Gilbert & Sullivan

Therefore, we have, in this simplest model, two classes of invariant solutions. With increasing complexity the maximal number of classes of invariant solutions is the degree of the polynomial. The condition of cloning an arbitrary qubit is, therefore, effectively satisfied even in the simplest model above with some phase coherence, and even better, I suspect, at high complexity. The resulting macroscopic phase-coherent quantum state has a local order parameter $\psi(x)$, which represents successful post-quantum cloning in the new sense of a coherent superposition of an uncertain number of clones. As the mean number of quantum clones increases to astronomically large finite numbers, the set of roots to the polynomial for the invariant unitary inner product of $|q'\rangle$ with $|q\rangle$ pre and post-cloning gets very large. The superfluid state, i.e. macro-cloned coherent state, with the local order parameter $\psi(x)$ [271] is of the form

---

[271] Think of x as an event in 4-dim spacetime. In general x will be an event in the full hyperspace of Super Cosmos.

$$|\psi\rangle = c_1|q\rangle + c_2|q\rangle|q\rangle + c_3|q\rangle|q\rangle|q\rangle + \ldots + c_n\underbrace{|q\rangle\ldots|q\rangle}_{n} + \ldots$$

The local order parameter is

$$\psi(x) \equiv \langle x|\psi\rangle$$

$$= c_1\langle x|q\rangle + c_2\langle x|q\rangle\langle x|q\rangle + c_3\langle x|q\rangle\langle x|q\rangle\langle x|q\rangle + \ldots + c_n\underbrace{\langle x|q\rangle\ldots\langle x|q\rangle}_{n} + \ldots$$

The "Bohm system point" material "hidden variable" or "particle" piloted by the qubit $|q\rangle$ "mental landscape" can be "real" on mass shell, or "virtual" off mass shell. The latter case is for the quantum superfluid vacuum that Haisch, Rueda and Puthoff have no notion of in their various models for "origin of inertia" and for "PV electro-gravitics" motivated by the UFO enigma.

Conclusion, Bose-Einstein condensates, especially non-equilibrium ones in pumped open systems like Frohlich electric dipole modes in biological membranes, are intrinsic post-quantum bit "Xerox" copy machines with increasing signal nonlocality as complexity increases. Therefore, what the no-cloning theorem really says is that it is impossible to use a reversible probability conserving "quantum gate" in a "quantum computer" to make a definite number of clones of an arbitrary input qubit |q>. Such a number eigenstate is completely incoherent. However, it is possible to make a partially coherent "superfluid" state of uncertain numbers of that arbitrary input bosonic qubit with long-range phase coherence in the amplified output. Indeed, this is what local order parameters in complex material systems are!

Spectral Associative Memories

"It does seem that causation from the future is becoming more mainstream. I was invited to present my work at a quantum computing workshop at NIPS2000 in Breckenridge (the work sparked by Jack's writings which led to 6 published papers and an integrated circuit)." by Ron Spencer, PhD, formerly at Department of Electrical Engineering, Texas A&M University where this work was done, now at "Silicon Laboratories in Austin now, just minutes south of Hal [Puthoff], working on coplanar wave guides and near-microwave optical transceivers."[272]

rspencer5@austin.rr.com
http://home.austin.rr.com/rspencer/
http://home.austin.rr.com/rspencer/Publications/
http://www.bbb.caltech.edu/compneuro/cneuro00/0112.html

---

[272] E-mail, December 19, 2001

Jack's New Idea for Space Force Engineers

Rules of Thumb for Star Gate Metric Engineering and Warped Mechanics ☺

Think of Einstein's geometrodynamic field equation like "Ohm's Law."

$$RI = V$$

R = resistance, i.e. a response coefficient

I = induced current, i.e. a response

V = applied voltage, i.e. a stimulus

R - > Super String Tension

I - > Einstein Curvature Tensor

V - > Stress- Energy Density Tensor

Einstein's 1915 GR field equation in modern parlance in 2002 is then

(Super String Tension)(Einstein Curvature Tensor) = (Stress-Energy Density Tensor)

Remember when Maxwell introduced the displacement current into Ampere's, Faraday's and Gauss's laws to get the electromagnetic wave explanation of light. Well that's the kind of thing I do here.

My "displacement current" analog is the partially cohered virtual quanta contribution to the Stress-Energy Density Tensor.

What Einstein did in 1915 in modern terms was only to include the real quanta (on the mass shell in special relativistic quantum field theory) that appear as "matter" in classical physics. There was no quantum physics in 1915 only Bohr's dim vision "through the glass darkly." Only Planck and Einstein's quantum particle momentum/wave number duality p = hk was then still a few years into the future.

225

Stress-Energy Density Tensor = Real Quanta Part + Virtual Quanta Vacuum Part

Einstein in 1915 only had the first term on the right hand side of this equation for obvious historical reasons.

The spacetime singularity theorems of Penrose and Hawking on black holes and big bans only assume the Real Quanta Part in their energy conditions like the NEC et-al.

In math language using Peacock's 3 GR sign conventions, the local field equation from 1915

(Super String Tension)(Einstein Curvature Tensor) = (Stress-Energy Density Tensor)

The simple obvious idea for warp drive and for star gate engineering is to soften the super string tension "resistance" for the fixed stress-energy density tensor. If we can't do that there is no hope and no truth to any of the alleged flying saucer observations as "nuts and bolts" machines. In that case we live in a boring universe where God is a clod, a Cosmic Dot.Commie, and it's Twilight of the Gods as the lights go out all over our accelerating universe where life has no far future. The alternative is the "O Brane New World" of Super Cosmos.

The above local field equation, neglecting irksome factors of 8π, in the 1915 limit, is

$$(\text{Super String Tension})G_{\mu\nu} = - T_{\mu\nu} \text{ (real)}$$

$$(\text{Super String Tension}) = c^4/G = hc/L_p^2$$

c = classical vacuum speed of light

G = Newton's constant of gravity

L_p = Planck distance i.e. lattice spacing of Hagen Kleinert's 4-dimensional world crystal lattice – a variable from 11-dimensional hyperspace in what Stephen Hawking calls "O Brane New World"!

How many 4-dimensional world crystal lattice discrete symmetry groups are there?

Classify the defects – how about 11-dimensional crystals?

Note the physical dimensions of Newton's G:

$$[G] = 1/(\text{density})(\text{time})^2$$

The "displacement current" analog for the partially cohered superfluid quantum vacuum is

$$t_{\mu\nu} \text{ (virtual)} = (hc/L_p*^2) \, \Lambda \, g_{\mu\nu}$$

Where the local cosmological variable Λ field is

$$\Lambda = (1/L_p*)^2[1 - L_p*^3 |\Psi|^2]$$

The local x-dependence is understood, x = 4 dim spacetime event. **Psi = Ψ** is the complex number spin 0 Higgs-like "mental" partial long range phase coherent quantum order parameter of the zero point fluctuations of all boson gauge force spin 1 fields plus the spin 2 geometrodynamic local $g_{\mu\nu}(x)$ field itself and all *fermion-paired* spin ½ and spin 3/2 source fields. The $g_{\mu\nu}$ field is the world crystal strain tensor from the Josephson gauge force invariant modulation of the local coherent phase field $\theta(x)$ of the complex function $\Psi(x) = |\Psi(x)| e^{i\theta(x)}$

$$g_{\mu\nu}(x) = \left(\frac{1}{2}\right)\left(\frac{\partial \xi_\mu(x)}{\partial x^\nu} + \frac{\partial \xi_\nu(x)}{\partial x^\mu}\right)$$

$$\xi_\mu(x) = \frac{\partial}{\partial x^\mu}\left[L_p^{*2}(x)\left(\theta(x) - \frac{e}{\hbar c}\oint A_\nu dx^\nu\right)\right]$$

Where $\xi_\mu(x)$ is the local infinitesimal distortion field of the world crystal corresponding to Einstein's local general coordinate transformations of the *topology-conserving* integrable (holonomic) Diff(4) group of curved spacetime. The gauge field line integral can, in principle, introduce retroactive *topology-changing* nonintegrable (anholonomic) corrections to Einstein's 1915 first approximation. The equation $\xi_\mu(x)$ is essentially the Josephson effect for the full Yang-Mills unified electroweak-strong force field of the standard model of lepto-quarks. It also generalizes the de Broglie-Bohm wave-particle "phase lock" equation from 3 space dimensions to 4 spacetime dimensions

$$\vec{v}(x) = \frac{\hbar}{m}\vec{\nabla}\left(\theta - \frac{e}{\hbar c}\oint A_\nu dx^\nu\right)$$

Note that the diffusion coefficient for the quantized fluid circulation and vorticity h/m becomes the variable Bekenstein c-bit $\sim L_p^*(x)^2$ *area* scale of quantum gravity information resolution, i.e. scale of the local concentration of unit cells of the world crystal.

The complete quantum corrected Einstein geometrodynamic local field equation is then

$$G_{\mu\nu} + \left(\frac{1}{L_p^{*2}}\right)\left(1 - L_p^{*3}|\Psi|^2\right)g_{\mu\nu} = -\frac{8\pi L_p^{*2}}{\hbar c}T_{\mu\nu}$$

This is the "It From Bit" equation. Einstein's 1915 limiting case is obviously $\Lambda \to 0$. Notice however that this requires the spontaneous broken quantum vacuum symmetry $\Psi \neq 0$! We need another "Bit From It" *back-action* Landau-Ginzburg equation to close the loop of spontaneous self-creation with inherent post-quantum signal nonlocality. We will also need a third hyperspace equation for $L_p^*(x)$ that I do not have yet in satisfactory form, but the theory is still young and the simple explanation of observed astrophysical results below justifies cautios optimism. It's not as though any of the Pundits have come up with an alternative idea to explain the enigma of the cosmological field, the dark matter, and the acceleration of the universe. In any case "Damn the torpedoes, full steam ahead!" The Landau-Ginzburg equation in classically curved spacetime for the Goldstone theorem spin 0 situation is of the form:

$$D^\mu D_\mu \Psi + \frac{\delta V(\Psi, \Psi^*)}{\delta \Psi^*} = 0$$

This Landau-Ginzburg equation is the Bose-Einstein condensate field theory analog of Newton's second law of particle mechanics in the Galilean relativity

$$\left(\frac{d}{dt}\right)^2 X + \frac{1}{m}\frac{dV(X)}{dX} = 0$$

$$F \equiv -\frac{dV(X)}{dX}$$

$$F = ma$$

in particle mechanics for topology conserving integrable "holonomic" reversible motion. In the analogy $\Psi \square X$ the displacement and $D^\mu D_\mu \square (d/dt)^2$ in one space dimension In the Landau-Ginzburg model with spontaneous broken symmetry

$$V(\Psi, \Psi*) = \alpha|\Psi|^2 + \beta|\Psi|^4$$

$$|\Psi|^2 \equiv \Psi*\Psi$$

$$\frac{\delta V(\Psi, \Psi*)}{\delta\Psi*} = \alpha\Psi + 2\beta|\Psi|^2\Psi$$

The local differential operator D_μ is covariant in both spacetime and in *internal space*, which is essentially hyperspace. Since, in this model Ψ is a spin 0 scalar field resembling the Higgs-Goldstone mechanism for spontaneous broken symmetry (as in my 1966 paper "The Goldstone Theorem and the Jahn-Teller Effect" with Marshall Stoneham in Proceedimgs of the Physical Society of London cited in the American Institute of Physics "Resource Letter on Symmetry in Physics")

$$D_\mu\Psi = \left(\frac{\partial}{\partial x^\mu} - \frac{e}{\hbar c}A_\mu\right)\Psi$$

$$D^\mu D_\mu\Psi = D^\mu\left(\frac{\partial}{\partial x^\mu} - \frac{e}{\hbar c}A_\mu\right)\Psi = \left(g^{\mu\nu}\frac{\partial}{\partial x^\nu} - \Gamma_\nu^{\nu\mu} - \frac{e}{\hbar c}g^{\mu\nu}A_\nu\right)\left(\frac{\partial}{\partial x^\mu} - \frac{e}{\hbar c}A_\mu\right)\Psi$$

$$= \left(g^{\mu\nu}\frac{\partial}{\partial x^\nu}\frac{\partial}{\partial x^\mu} - g^{\mu\nu}\frac{e}{\hbar c}\frac{\partial}{\partial x^\nu}A_\mu - \Gamma_\nu^{\nu\mu}\frac{\partial}{\partial x^\mu} + \Gamma_\nu^{\nu\mu}\frac{e}{\hbar c}A_\mu - \frac{e}{\hbar c}g^{\mu\nu}A_\nu\frac{\partial}{\partial x^\mu} + \left(\frac{e}{\hbar c}\right)^2 g^{\mu\nu}A_\nu A_\mu\right)\Psi$$

Summing, of course, over repeated tensor dummy indices in computing the totally covariant Helholtz wave operator $D^\mu D_\mu$ in curved spacetime with Yang-Mills gauge force potentials A_μ. Therefore, in general we can get the "Mexican

229

Hat Potential" http://www.sigmaxi.org/amsci/articles/00articles/ganguicap4.html giving $\Psi \neq 0$ from an "ODLRO" quantum vacuum phase transition into a Bose-Einstein condensate of virtual bosons and virtual fermion/antifermion pairs all of the mass shell of their respective Feyman propagators.

The completely incoherent zpf quantum vacuum of Haisch, Puthoff and Rueda corresponds to the "slice"

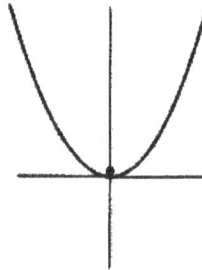

$$V(\Psi), \alpha > 0, \beta > 0 \text{ for } \Psi = 0$$
Haisch, Puthoff & Rueda SED Model

This phase of quantum vacuum does not even permit the classical Einstein geometrodynamic field of curved spacetime $g_{\mu\nu}(x)$ to come into being because $g_{\mu\nu}(x)$ comes from the modulation of the *long range* "Josephson Effect" macroscopic phase coherence from the spontaneous broken symmetry *local* order parameter $\Psi(x)$. Thus, Haisch, Putthoff and Rueda are not asking the right question in their vain quest to implement Andre Sakharov's vision of 1967 to derive Einstein's geometrodynamics from the quantum zero point fluctuations. In fact there is no consistent incoherent classical limit to quantum gravity as is required by the "stochastic electrodynamics" or "SED" idea promoted by Haisch, Puthoff and Rueda.

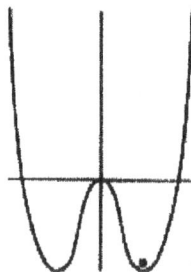

$$V(\Psi), \alpha < 0, \beta > 0 \text{ for } \Psi \neq 0$$

Required for classical curved spacetime to come into being in Jack's Model.

This phase transition has the critical point $\alpha = 0$ inside the grand quantum vacuum of all the actual fields results in a "holographic" local partial coherence order parameter $\Psi(x)$ with long-range quantum phase coherence in which Einstein's classical geometrodynamic field $g_{\mu\nu}(x)$ arises like Botticelli's Venus from the Sea in a self-organizing self-consistent bootstrap in which "More is different" in the sense of P W Anderson.

Remember the whole idea of warp drive is that the alleged hypothetically neutral "flying saucer" is creating and steering the free float weightless timelike geodesic at its center of mass with small tidal forces over the scale of the craft and with small expenditures of $T_{\mu\nu}$ (Real Quanta). Nothing less than that will do. Nothing less than that counts as "propellantless propulsion." The kind of "electromagnetic stress propulsion" that James Corum, ex-chief scientist at SARA in Huntington Beach, is hoping to get at the Institute for Software Research in Senator Robert Byrd's neck of woods, is not good enough because, even if achieved, the center of mass of their "flying saucer" will not be on a free float weightless timelike geodesic as in the Alcubierre warp drive math model, for example. Our alleged "Black Ops" crew (of the conspirational lunatic false urban legends in the Internet and in "UFO Disclosure") will free g-force just like

I felt when catapulted off CV 61 in an S3 in 1987 in the Indian Ocean. That is not good enough for the Space Force.

Jack reviews *The Fleet, 1987* ☺

What about the "torsion" of the Moscow "SDI" School in Russia like Gennady Shipov, for example? Gnnady was my guest at ISSO for an extended time. I will treat this issue in my second book "Super Cosmos" Everything here is done in the standard 1915 franework with zero torsion and with metricity. This means that the spacetime connection field $\Gamma^{\lambda}_{\mu\nu}(x)$ is symmetry in the two lower indices and that the simple Bianchi identity holds leading to vanishing covariant divergence of the Einstein $G_{\mu\nu}(x)$ tensor field and also metricity is vanishing covariant divergence of the metric tensor field $g_{\mu\nu}(x)$. That is,

$$D^{\mu}G_{\mu\nu}(x) = 0$$
$$D^{\mu}g_{\mu\nu}(x) = 0$$

In Einstein's 1915 theory with zero torsion and zero $\Lambda(x)$ field from partially cohered virtual vacuum zpf quanta, at fixed Newton's G, the above two conditions give the local conservation of what John Archibald Wheeler calls "momenergy" ("A Journey into Gravity and Spacetime"), i.e.

$$D^\mu T_{\mu\nu}(x) = 0$$

This is no longer the case with my new equations that might even get more complicated with Shipov's torsion field thrown in.

$$G_{\mu\nu}(x) + \left(\frac{1}{L_p^{*2}(x)}\right)\left(1 - L_p^{*3}(x)|\Psi(x)|^2\right)g_{\mu\nu}(x) = -\frac{8\pi L_p^{*2}(x)}{\hbar c}T_{\mu\nu}(x)$$

$$D^\mu G_{\mu\nu}(x) + D^\mu\left\{\left(\frac{1}{L_p^{*2}(x)}\right)\left(1 - L_p^{*3}(x)|\Psi(x)|^2\right)g_{\mu\nu}(x)\right\} = -D^\mu\left\{\frac{8\pi L_p^{*2}(x)}{\hbar c}T_{\mu\nu}(x)\right\}$$

$$g_{\mu\nu}(x)D^\mu\left\{\left(\frac{1}{L_p^{*2}(x)}\right)\left(1 - L_p^{*3}(x)|\Psi(x)|^2\right)\right\} = -\frac{8\pi L_p^{*2}(x)}{\hbar c}D^\mu T_{\mu\nu}(x) - T_{\mu\nu}(x)\frac{8\pi\frac{\partial}{\partial x^\mu}L_p^{*2}(x)}{\hbar c}$$

Obviously, we do not need the Haisch-Puthoff-Rueda completely incoherent zpf SED ansatz or the Puthoff PV "dielectric" ansatz, or Yilmaz local classical gravity energy, or any off beat ideas to do anything interesting at all. In fact in terms of practical results and real insights they are dead ends around for more than a decade with nothing much to show that I can see.

We do need Einstein 100%, we do need Kleinert http://www.physik.fu-berlin.de/~kleinert, we do need tensor mathematics, we do need the big bang, we do need the universe's unseen dimensions in hyperspace, we do need mainstream physics especially ideas of spontaneous broken symmetry "more different" (P.W. Anderson), we do need "the warped dimensions" (an eccentric term coined by Joe Firmage in his New Age pep talks on the physics of "propellantless propulsion" of UFOs eschewing Einstein). Indeed, we need everything the Motion Sciences Group http://www.calphysics.org, http://www.motion-sciences.org from The Academy of Laputa (J. Swift's "Gulliver's Travel") wants to throw away! This Academy is not the same as "The Invisible College" mentioned by Jacques Vallee.

Note that the equation of state of the zpf (virtual zero point fluctuasion) for all quantum field is

$$\text{zpf energy density} + \text{zpf quantum pressure} = 0$$

Ref. Peacock "Cosmological Physics" pp 25-26 Cambridge Press, 1999.

Poisson's equation of gravity in the Newtonian weak curvature limit for the quantum vacuum as a relevastic fluid is

$$\nabla^2 U(\text{virtual}) = -(4\pi G/c^2)(\text{zpf energy density} + 3\text{zpf quantum pressure})$$

Note the all-important factor of 3.

U(virtual) = **quantum vacuum gravity potential energy per unit mass**

$$[\nabla^2 U] = 1/(\text{time})^2$$

$$[G] = 1/(\text{density})(\text{time})^2$$

Therefore,

$$\nabla^2 U(\text{virtual}) = (8\pi G/c^2) \, (\text{zpf energy density})$$

but

$$\text{zpf energy density} = t_{00}(\text{virtual}) = (hc/L_p{}^{*2})\Lambda g_{00} = (c^4/G) \, \Lambda$$

Therefore

$$\nabla^2 U(\text{virtual}) = 8\pi c^2 \, \Lambda$$

The dimension on both sides of Poisson's equation for the quantum vacuum gravity are $1/(\text{time})^2$. The Peacock sign conventions are such that

$\Lambda > 0$ antigravities as "quintessence" (speeding up of expansion rate of our big bang universe)

$\Lambda > 0$ gravities as "dark matter" (~90% of the effective "mass" in the universe)

Einstein's 1915 classical geometrodynamics has $\Lambda > 0$ because there are no virtual quanta Bose-Einstein condensates off the mass shell in that early model.

Einstein called $\wedge$ his "greater blunder." In fact, it may turn out to be his greatest triumph. Ed Witten said that superstring theory is 21st Century Physics that fell into the 20th Century. Well it's now 2002 AD.

Ad Astra and Beyond to Super Cosmos
Sagan's Cosmos is only a small piece of Super Cosmos
Sagan's Cosmos with Hawking's chronology protection is Flatland
Stale and unprofitable
O Brane New Worlds!

Indeed, a proper title for a mainstream book on cosmological physics with Hawking's chronology protection conjecture forbidding time travel to the past in an accelerating universe would be:

"The Boring Universe" ☺

The lights would not only blink out all over Europe, but, indeed, all over Sagan's Cosmos in Gotterdamerung.

http://www.pnc.com.au/~marlan/wagner/gotterdammerung.htm

"You now face a new world, a world of change. The thrust into outer space of the satellite, spheres, and missiles marks a beginning of another epoch in the long story of mankind. In the five or more billions of years the scientists tell us it has taken to form the earth, in the three or more billion years of development of the human race, there has never been a more abrupt or staggering evolution. We deal now, not with things of this world alone, but with the illimitable distances and as yet unfathomed mysteries of the universe. We are reaching out for a new and boundless frontier. We speak in strange terms of harnessing the cosmic energy, of making winds and tides work for us, of creating unheard of synthetic materials to supplement or even replace our old standard basics; to purify sea water for our drink; of mining ocean floor for new fields of wealth and food; of disease preventives to expand life into the hundreds of years; of controlling the weather for a more equitable distribution of heat and cold, of rain and shine; of spaceships to the moon; of the primary target in war, no longer limited to the armed forces of an enemy, but instead to include his civil populations; of ultimate conflict between a united human race and the sinister forces of some other planetary galaxy; of such dreams and fantasies as to make life the most exciting of all times." General Douglas Mac Arthur, "Duty, Honor, Country", West Point, 1962 http://www.ohiomp.org/thayer.html

My Spiritual Path by Hyman Sarfatti [273]

"There is no creature who is not destined for the supreme goal as there is no river which is not winding its way to the sea." - Avatar Meher Baba

Hyman Sarfatti at Age 87 and Jack Sarfatti

Born in The Cave 1914

Life is the only literature that lives and if I had not first lived this story, it would not be worth the writing. This consciousness is sufficient recompense for relating what has made an indelible impression on my mind.

As related to me by my older sister, in 1914, the year in which I was born, a war had been raging for two years between Bulgaria and Turkey. Unfortunately, Greece became a victim of the hostilities because the armies of both countries would cut across Greece, north to south and vice-versa. The town of Monastir, a suburb of Kastoria in which I was born, was in the left-center of Greece and

---

[273] Jack Sarfatti's 88-year-old father.

237

made a direct passage-way for the two opposing armies to cross in order to get at each others front lines.[274]

The Bulgarians, as it turned out, were a good and principled people because as their armies would cut across Greece, north to south to get at the front lines of the Turks, they did not harm the Greek population in any manner whatsoever. In fact, if they took anything they needed such as food, drink, fodder for their horses etc., they would pay for everything they needed.

However, the Turks were a brutal and bloodthirsty people who took what they needed, never paying for anything. In fact, they would rape the women and kill anyone who resisted, male or female. For obvious reasons, lookouts would be posted in the hills so that when Turkish troops would be spotted advancing, the lookouts would jump on their donkeys and ride through the town yelling, "The Turks are coming! The Turks are coming!" This is reminiscent of Paul Revere in our Revolutionary War riding his horse yelling, "The British are coming! The British are coming!"

Of course, the American people's army made their stand and battled the British, but the Greek people simply gathered up their children, packed everyone's back with everything they could carry for survival and headed for the hills in which there were many caves.[275] They would live in these caves until all the Turkish troops had passed and it would be safe to go back to their homes.

As Fate would have it, my very dear and wonderful mother was pregnant with me and on one of these occasions, the hurried and forced march into the hills brought on birth pangs and I was born in one of these cold, damp and dark caves in the winter of November 1914. The midwives simply stretched out a blanket on the rough earthen floor and I was born under candlelight, as my good mother related it to me.

It is ironic that the only other Jewish individual I know of that was born in a cave was Jesus.

However, Mary gave birth to Jesus in a cave that had been converted into a stable to house the animals that came into town with their owners, like we park our cars in garages or parking lots. The stables had stalls called mangers filled with plenty of hay and lit up with oil lamps. Besides, there was an inn next door where all amenities could be purchased and there were no cruel and murderous enemies at the "door". Therefore, I may say that Jesus was born in a Hilton Hotel compared to where I was born, and my mother had a much harder time, generally speaking, than Mary did.

Strangely enough, I have always had a heart-felt sympathetic affinity to Jesus even as a teenager. My sister Regina, the oldest in the family often told me the

---

[274] Hyman's father was from Naples, Italy. His business was trade in fruit between Italy and Greece.

[275] Not unlike Afghanistan today.

story that when I was a baby in the crib, I would wake up at night crying and asking for water in Ladino, "Yo quiero agua - Yo quiero agua!" "I want water, I want water!" My mother had assigned sister Regina the duty of caring for me in the night time so that mother could get a well-deserved and needed rest at night as she worked in the day-time taking care of the family.

Sister Regina would bring me a cup of water to drink and say, "Here is your water," in Ladino[276], but I would shake my head saying, "No, yo quiero agua, no yo quiero agua!" My sister would get exasperated repeating, "Here is your water," in Ladino and I would repeat, "No, no quiero agua, no yo quiero agua." This went on for months.

I remember this story well because when I grew older, I realized intuitively that the child in the crib was asking for the spiritual water of Truth, not the drinking water. This has been the key signature to my life, for I have always been searching for the Truth.

Beat The Devil

When I was about seven or eight years old, I had periodically, the most frightening of dreams. I dreamt that I was in the bottom pit of Hell and seeing the Devil clearly in all of his ugly fierceness scowling at me. I was scared out of my wits, but slowly I felt a strong and distinct surge of courage well up in me coursing through my body and into my heart. Then I would stand tall facing the Devil, look straight at him and say, "I'm not afraid of you; God will protect me. I'm not afraid of you; God will protect me. Do your worst, I'm not afraid of you, God will protect me." I remember to this day the exact words I voiced in defiance. I said those words with such a bravura that the Devil seemed to cringe and slowly fade away. Our dream world is just as real as our awake world, each having its own reality and running the gamut of fears, doubts, worries, despair, resentment, anger, hatred, hope, courage, pains and pleasures, happiness and misery, all seeming real in both the dream and the awake world.

I had this same dream periodically for many months where I got to the point where I dreaded going to sleep for fear of having this most frightening dream recur. But who can give up sleep? So reluctantly, I would lay down in my bed and think, "Oh well, if I should have the same dream again, God will protect me. I had a simple child-like faith in God that was instinctive and natural. I did not have to think about God. God was as natural as flowers and trees, birds and stars. Around that time, a boy my age living in the same building and whose

---

[276] 15th Century Castilian Spanish mixed with Hebrew from 1492 when Ferdinand and Isabella expelled the Jews from Spain because they were the "brains" of the Moorish Empire. Ladino is to Spanish as Yiddish is to German.

father had recently passed away said to me, "My mother says that my father is in Heaven with God and that he is an angel." Then he looked up in the sky pointing and saying, "See those angels in the sky." I would say, "Where, where?" I didn't doubt him for a moment, only I couldn't see the angels. Thereafter, for many days I would periodically look up in the sky hoping to see angels. Of course there were angels but I couldn't see them just yet. Who could doubt God and His angels?

Along about the same time when I was seven, eight and nine, there were some ladies, all grandmothers who all knew each other, who would come to visit my mother to chat over a cup of demitasse Turkish coffee of the most wonderful aroma and taste. I would ask my mother for some, which she always gave me and I would stand aside sipping my coffee. I enjoyed listening to these wonderful ladies conversing in Ladino, which I understood quite well. I was struck by the fact that in practically every sentence spoken, God was mentioned. The phrase used was, "El Dio Santo," the Holy God. It was like a golden thread running throughout the fabric of their conversation. I felt very good listening to them. I really enjoyed their 'company.' This made a deep impression on my young mind and perhaps helped me in my dreams of Hell and the Devil, which occurred around the same time.

Sometimes one, two or three of these ladies would drop by when mother was not at home. I would welcome them in, as I was always delighted with their company. Immediately, I would take out the real, traditional, copper Turkish coffee pot that my father had told me that they had brought with them on the boat from Greece when they sailed to the United States. I would immediately start making some Turkish demitasse coffee that I learned to make by watching my mother make it innumerable times. The ladies would tell me not to bother, but I would insist that they stay and have some coffee. I was proud and happy to serve them and they seemed pleased with my hospitality. They told me how good the coffee was and praised me for it. I simply adored these ladies and always felt happy in their presence.

There was one lady that I remember in particular, Tia Luna, who seemed to be the conversational leader of the group. I recall clearly how she said that she fasted one day a week every week of the year, including of course Yom Kippur. It was a religious fast in which she always prayed to God, El Dio Santo, asking for His help for some sick individual or family member having some difficulty. She was indeed a Saint and I remember her like it was only yesterday.

In touching on the subject of fasting, let me say that it is a common denominator for every religion in every part of the world. It is a part of religious observances such as the one-day Fast of Yom Kippur of the Jews or the Thirty-Day Fast of Ramadan of the Moslems. Both are exactly alike except that the Moslems fast for thirty days in a row or thirty Yom Kippurs consecutively. The relationship between fasting and religion is so pervasive that one cannot separate them. They are heads and tails of the same coin.

Mahatma Gandhi said, "I know that there is no prayer without fasting and all fasting, if it is a spiritual act, becomes an intense prayer. I can as well do without my eyes as without fasts. What the eyes are to this outer world, fasts are for the inner world.

St. Clement (AD 150-215) a Greek theologian said, "Fasting is better than prayer." The prevalent beliefs of the ancient Greeks was that the soul reached its greatest power when totally cut off from the lower digestive system of the body and relinquishing sex-activity as well. The fast, no food and no sex, freed the soul from the influence of the senses of the body and brought the individual nearer to the Gods and receive their wise messages. The Romans as well as the Greeks believed that through fasting one could become an oracle, a medium through which the gods could speak and communicate their wisdom.

It is a said of Pythagoras, a Greek philosopher and mathematician (582-500 BC), that his fasting practices generated the insights for his most elegant mathematical theories and helped develop his scientific skills and power of prophesy. Also, one of the basic tenets of Pythagorean philosophy is the belief in reincarnation.

During the early Christian era men fasted as an ascetic act that included rigorous self-discipline, self-denial, silence and prayer, thereby attaining virtues.

Moses, Buddha and Jesus, all three fasted for forty days and forty nights completely, unlike the Fast of Yom Kippur or the Thirty-Day Fast of Ramadan when food is consumed once every twenty-four hours.

Moses received the Ten Commandments from God after his forty-day complete fast on Mount Sinai.

Buddha attained Enlightenment after his forty-day complete fast under a banyan tree. Then he went on his mission teaching his Eightfold Path to Perfection and his Wheel of Rebirth, to live eternally in Eternal Life with God in full consciousness. Let me add that the face of the Buddha as represented in statues of the Buddha is an expression of the calmness of the Cosmic Consciousness or the state of Nirvana.

The philosophy of Buddhism states:

The Four Noble Truths

1. The existence of suffering
2. The cause of suffering (ignorance)
3. The cessation of suffering
4. The Eightfold Path to Perfection that leads to the cessation of suffering

The Eightfold Path to Perfection

1. Right belief
2. Right thought
3. Right speech
4. Right living
5. Right action
6. Right work
7. Right meditation
8. Right adoration of God

Jesus fasted for forty days and nights in the wilderness and was tempted by Satan to give up his faith in God, following His precepts and doing His Will. In return, Satan would give him unlimited wealth and worldly power. It is ironic that both Jesus and I were born in a cave in the humblest of circumstances and we both wrestled with Satan, Jesus in his way as an adult and I in my way as a young boy. And we both emerged victorious. God always wins, never loses, so it behooves us all to stay with God. It is never too early or too late to be spiritually minded and have supreme faith in and love for God. By our own thoughts, words and deeds, we make our own Heaven and our own Hell.

When I was in my twenties, the thought occurred to me that if I had a choice of being any famous or prominent individual in all of history in any major field of endeavor, who would I choose to be like the most. The answer came to me instantaneously. It was Jesus. To me he was the greatest—Numero Uno. Jesus was the sweetest, sanest and dearest individual who had ever walked on the face of the earth. I have always had the greatest respect and love for Jesus, even though I was raised in the Jewish tradition, went to Hebrew school, attended a Sephardic Synagogue which I loved, was circumcised and Bar-Mitzvah'd at thirteen. I was a 'Jew for Jesus' long before the Jewish individual who started the organization known as 'Jews for Jesus' was born, or perhaps he was still in his diapers. This organization was started in San Francisco and its headquarters are still here.

When I was about ten, eleven and twelve, I remember living on the lower East Side of Manhattan and that I would often walk across town to get to the Italian section. There the most wonderful fruit and vegetables among other foods were sold on pushcarts and storefronts. It brought joy to my heart to see crowds of people happily shopping. The activity was festive and I would buy a piece of delicious fruit and ate it as I sauntered along. It was a pleasant way to spend the afternoon.

Now, to get to the Italian section from where I lived, I had to cross the Bowery, which was then the Mecca for alcoholics. Near to the intersection where I would cross was a Salvation Army chapter, a storefront with a deep interior. As I looked inside through the storefront plate glass, I could see all the way forward a pulpit with chairs in front of it, sometimes occupied. Obviously, a Salvation Army chaplain would give a sermon and then after would serve sandwiches, donuts and coffee. In short, first food for the soul then food for the body is a practical and wise combination in the right order. This tradition and service has always engendered in me a great respect and love for the Salvation Army.

Now in the facility up close to the plate glass was a huge Bible on a bookstand similar in size to the large Webster's dictionary one may find in a library on a stand. I would read the open pages, and it seemed to me that every time I passed by, the Bible would be turned to another page and another chapter. It was then and there that I started reading the New Testament. The Bible also contained the Old Testament but more often it was open to the New Testament. As young as I was, I was very literate and could read the Bible with ease and could understand it well. I did not find any conflict with the Old Testament. The Five Books of Moses and the New Testament were to me the repositories of the most and best Truths. My intuitive heart told me it was the truth from God. All religions come from God and are meant to bring people back to God. I certainly had no problem understanding the teachings of Jesus, which were to me as clear as crystal. I have always enjoyed reading the Scriptures of all religions. The reading of Scriptures develops a devotional bent of mind and makes one humble. Before God, one cannot become an egotist or ostentatious.

I started high school at the age of twelve. I was generally a good student and well behaved. One year I had a wonderful English teacher who taught us well. She focused upon and insisted that we learn how to read, write and speak English with excellence. Her favorite homework was to read some classic in literature and write a book-report on it. In class daily, she would assign a student to write on the black board about some character or event in the book that we were studying. Every student took a turn at the blackboard. Of course my turn would come. I saw behind the events and characters in the story a meaning of moral and spiritual significance. Things did not just happen. There was clearly cause and effect, or "As ye sow, so shall ye reap" or the Law of Compensation or Retribution, or Karma as the Hindus call it. Invariably, in writing my composition on the blackboard, I would inject some quote from the New Testament attributed to Jesus, which would explain from the moral point of view, why things happened as they did. I knew that these events and results were not manifestations of some blind, foreign or oppressive Fate. They were based on the inexorable spiritual laws of God in Creation. The Scriptures made that clear.

I knew my compositions were flawless as I could read and write comparably and equally with any college student, by today's standards - perhaps better.

We were studying Ivanhoe when it came my turn at the blackboard. I chose to write about the wicked Black Knight, his evil character and wicked ways and the final results of all of his wickedness. He had an insatiable appetite for wealth and power.

His selfish ego-mind dominated all of his thoughts, words and actions. I injected the teachings of Jesus to explain why he ended up as he did, as just retribution for his demoniac character and wicked actions. After finishing writing our composition on the blackboard, we were required to read it out loud before the teacher and classmates.

From my opening sentence, the class went into such a silence that you could literally hear a pin drop. I had my student audience mesmerized as they concentrated on every word I spoke. When I came to the end, their reaction was as if they had just gotten out of a spell and remained speechless momentarily. Moments later, my English teacher asked me how come I knew so much about Jesus. I cannot forget her exact words even to this day. She said "Hyman, you are a nice Jewish boy; how come you know so much about Jesus? I answered simply, that I liked reading the Bible. Thereafter, whenever I wrote a composition to be read out loud, absolute silence was the norm. A student or two would even shush the class at the slightest disturbance. When Hyman spoke, everyone would listen.

The subtle power that I exhibited unquestionably came from the spirit of the Truths that I had gleaned from the Bible. Words come alive and are potent when the Spirit of Truth is behind them. I had a facility for reading and understanding Scripture with remarkable ease. Nothing was too deep for my understanding. I have been told more than once that I should have been a preacher.

Let me add that my English teacher always gave me two tens for a single composition, a ten equaling 100% - two tens were the equivalent of 200%. Perhaps one ten was for flawless composition and other ten for scriptural knowledge. I am probably the only student in any English class that rated 200% for a single composition.

One day in the spring, I was walking on the Boardwalk of Coney Island. I was about seventeen at the time, when I came upon an intersection where the street ran up to the boardwalk. At the corner right off the boardwalk, I saw a large individual standing on a wooden portable pulpit with a large book in his hand, preaching. I knew immediately that it was a Christian evangelist, because such sights were common in those days. We had our Billy Graham's all over the place. I approached and joined a small crowd listening with rapt attention. I focused on every word he uttered with a booming voice and very articulate presentation and with great sincerity. He was telling the crowd of the events preceding the Crucifixion of Jesus and the Crucifixion itself. His speech was so moving and I felt so sorry for the suffering Jesus that tears started welling up in my eyes and rolling down my cheeks. I didn't realize it until a cool breeze from the ocean rolled in and cooled my warm tears. Realizing suddenly that I had been weeping I looked around me furtively to see if anyone else had tears in his eyes. I saw that I was the only one and suddenly the thought came to me, "I am a Jewish boy and Jews are not supposed to cry over Jesus."

Immediately, I took out my handkerchief and gently blew my nose, feigning that my tears were only due to a cold. I felt somewhat embarrassed that I had wept for Jesus. Can you for a moment, picture a Jew-boy in a crowd of Christians weeping for Jesus? That preacher could have made a Sphinx melt into tears. My tears were totally sincere and honest. By then I had a very sympathetic rapport with Jesus as we're on the same 'wavelength'.

Remaining on the subject, I was in my early twenties when one day in the morning, I had gone to the Main Public Library on 42nd Street and Fifth Avenue. Looking through the catalogue on Religion and Philosophy I came across a title that intrigued me. It was `The Life of Jesus the Christ' by an author named Levi. I took it out immediately to read it in the main reading room. I remember that it was 9:30 am when I started reading the book. It was so interesting that around lunchtime, I debated with myself whether I should stop and go out for lunch or continue reading. I decided that the book was much too interesting to stop even for lunch, so I went to the men's room, had a drink at a water fountain readily accessible and then went back to my place and continued reading. I never ate breakfast with perhaps only some acid fruit juice, so that I fasted all day while reading.

Amazingly, I focused on the book with such concentration that I was aware of my mind but not my body. I was all mind and no body unaware of my surroundings. I kept reading the book until five minutes before 10pm, when a bell would ring announcing the closing of the Library.

I got up and was reading the last page as I walked up to the return counter. I had finished this version of the New Testament, cover to cover in twelve and one half hours of total concentration. I felt as light as a feather as if walking on a cushion of air. My mind felt clear and pure, my heart was buoyant with a sense of emotional equilibrium, and my soul cleansed of any residue of impurity. I felt happy like nothing in the world could disturb me. The sky could have fallen down at the time and I would not have been the least bit upset. It may be said that I got a `Spiritual High.'

I felt generally good the following week when I thought of my wonderful experience. And so I decided to go back the Main Library and read the same book again. I got the book out by 9:30 am and reread it. At around noontime, I got up to go the men's room, drink some water at the fountain and returned immediately to the reading room. The thought of having lunch did not appeal to me and seemed like a sheer waste of time when I had so much mental and spiritual 'food'. The call of my soul for the Truth overrode the call of my body for food. I read the Aquarian Gospel of Jesus the Christ' from cover to cover with total concentration, finishing by five minutes to 10:00pm. Again, I experienced the same 'spiritual high' as before.

A week later, I got the urge once more the go to the Main Library and read the book once more cover to cover. I finished reading the book in time for the closing of the Library. This was the third reading in two weeks. I got the same 'spiritual high' for a third time.

This can be compared in a manner to the 'high' gotten from drug-use, with a big difference. The 'high' gotten from drug-use has a bad after effect, resulting in depression, misery, pain and illness. The 'spiritual high' brings health, joy, mental stability, emotional equilibrium, courage, faith and hope - with no unwelcome side effects.

Drug use develops rapidly into a craving for the pleasure it gives followed by pain in an endless cycle of pleasure and pain. In my 'spiritual high' I was likewise developing a craving or addiction - but for God, Jesus and Truth, with no pain or unwelcome side effects. In fact, to become addicted to God frees one from all the dualities of a materialistic life, such as pleasure and pain, happiness and misery, health and sickness, success and failure, riches and poverty, etc.

On the peak of the God-Consciousness or Cosmic Consciousness storms never beat, misfortunes never settle, vicissitudes never venture and disasters never fall, because that individual has risen above all mortal ignorance, weakness and sin. The wisest thing that anyone can do is to develop a predisposition for God, Truth and Love. Nothing is real but God, because God is permanent or immortal and He never changes. Everything else in the Universe is in a process of change through the dual aspects of Evolution and Involution (the development of consciousness.) Therefore, everything in the Universe has only a temporary or relative reality. Therefore, the whole Universe is not real—only God is Real.

Speaking of libraries, when I was about eighteen I visited a local library and was looking in the section marked Philosophy and Religion when my finger touched a book, as I was running it across the various titles. The book fell to the floor, face-up and I read the title, The Bhagavad Gita. The title intrigued me, so I picked it up and went to the reading room. After about a half-hour, I said to myself, "Wow, how long has this book been around." At the time I did not realize that it was the Numero Uno of the many Hindu Scriptures. The book is an actual conversation between Krishna an incarnation of God, about five thousand years ago with Arjuna, His friend, companion and disciple. God as Krishna dictated the Bhagavad Gita to his top disciple, Vasudeva, and it is truly God speaking directly. I read the Scripture through and through and reread it two or three more times. Then I decided to write a short version of it, taking excerpts that seemed to pertain to my understanding and me.

This copy, written in longhand on white blank paper, I carried with me in my back pocket, reading and re-reading it countless times. It always had a beneficial effect on me. It seemed to lift my spirits and gave me hope for the future. I treasured it as much as I did the teachings of Jesus. I regarded them both as being on the same level.

Again, when I was about eighteen, I joined the Rosicrucian Order, headquarters in and received weekly manuscripts. The basic study is of all religions and philosophies and to bring one to God and become ultimately Self-Realized or One with God. I read all the books published by the Order and read many of their monthly issues of a magazine called the Rosicrucian Digest. Among the dozen or so books published were two that I loved to read, (a) The Mystical Life of Jesus (b) The Secret Doctrines of Jesus. I was already familiar with the teaching of Jesus through the New Testament. By the time I read The Aquarian Gospel of Jesus the Christ, I was convinced that Jesus was the Messiah that the Jews never recognized even to this day. I was indeed a 'Jesus freak' on the highest intellectual level. In fact, I believe that I understand Jesus better than the Pope himself.

At this point, let me go back to when I was twelve years old. My older brother Sam had bought a set of the Bernarr McFadden Encyclopedias of Physical Culture. Bernarr McFadden was then the nationally famous Guru of health and Physical Culture. He also published a monthly magazine called, Physical Culture which was very successful and popular.

My brother Sam practiced the lessons for a time and then gave up. He told me he was too busy to continue his efforts and I could have the set of eight volumes. I was very literate at twelve and could read the encyclopedias with ease and understand everything. I was fascinated by its teachings and started immediately to follow his system of health building, which included physical exercises, diet which was most important, yoga, breathing methods, all kinds of baths and exposure to air, sun and earth, massage and aids to health and curing diseases with natural methods, including fasting.

In four years of steady practice and adopting a strict vegetarian diet, which was recommended as the ideal diet for building and maintaining health of body and mind, I achieved perfect health, incredible stamina, great strength and a beautiful body. I may interject that Yoga Ashrams around the world serve strict vegetarian menus and Yogis on the spiritual path are strict vegetarians to a fault. Back to Nature and up to God is the essence of their philosophy.

As to vegetarianism, Appolonius of Tyana (A.D. 10-98) a Greek philosopher said that he refused to eat meat on the grounds that it defiled the mind and rendered it impure. He considered that the only pure food was what the earth produces, fruits, vegetables and nuts. He also abstained from wine, because it disturbed the composure of the mind. He was 100% vegetarian, like all Yogi's are. Yoga philosophy says that meat has vibrations not conducive to health and spirituality and which excites the passions of lust and anger. A vegetarian diet, they say (Yogis) calms the mind and the passions are greatly reduced, make one more self-controlled and is conducive to spirituality. The mind actually becomes clearer and stronger. They are absolutely right because I have proved all of this myself in the practice of vegetarianism over long periods of time.

Jack Sarfatti

Appolonius of Tyana also believed in re-incarnation and even spoke of who he was in a previous incarnation.

Plotinus, a Greek philosopher in the same era (AD 204-270) thought along the same lines and emphasized the unity of all creation and the ideal of oneness with God-God, Nature and Man or unity Consciousness and no real separation between the three.

Let me relate a significant incident. In my neighborhood lived a young individual, my age of sixteen, who was a bully who terrified the whole neighborhood, by challenging anyone and everyone, teen-agers and adult alike, to engage in fighting him.

One day I was playing handball, by myself when suddenly this bully came upon the scene and grabbed my ball. He started playing with it as I had been doing, when after a few minutes, I asked for my ball back. He said that if I wanted my ball back I would have to take if from him.

He was provoking me, as was his habit, to fight him to get my ball back. He would use any ruse to get someone to fight him. I accepted his challenge. As we squared off like two fighters in a boxing ring, he began weaving in and out, side to side like a professional boxer. I waited for him to come in and throw a punch. As he came in close to throw a punch, I threw one with lightning speed, so fast he never saw it coming. I struck him such a powerful blow that he went reeling backwards and fell like a board of wood. He looked so surprised when he got up and said, "You win" with a wan smile and stretched out his hand to shake mine. One blow struck and the fight was over. It was my left hand. Had I been able to follow up with my right hand, he would have needed a medic to revive him.

I was in top shape physically and he was unaware that I had trained myself for four years steady and had remarkable strength and stamina for my age. Thereafter, when I would run into him occasionally in the neighborhood, I was struck by the fact that his whole personality had changed. He acted sweet and gentle with a ready smile and without the slightest trace of arrogance, belligerence or prone to violence. He even said once that if I ever need help, to call on him. He became a friend instead of a potential enemy. I believe that my purity of body and mind and my proneness to spirituality and religious bent of mind, got to him subtly and he was 'reborn' in the moral and spiritual sense. I say this in retrospect. We influence others by our own state of body, mind, heart and soul for good or ill.

In my mid-twenties, World War Two had started. I went to trade school and learned enough to get a job as a ship fitter, 3rd class in the shipyards of Bethlehem Steel in Brooklyn. For some obscure reason I transferred to a job at Todd Shipyards In Hoboken, NJ and I was upgraded to ship fitter 2nd class. I enjoyed my work and the pay was good for those days. After about four years through the war, working six days per week ten hours a day, I contracted a cold and cough that I neglected for weeks. Besides, I was smoking cigarettes, which was bad for me. In those days the cancer causing potential of smoking was not in the public consciousness at all. In fact, it was unknown and unspoken. Let me add that at some time, I was notified by the Draft Board, to come down and be examined by government doctors as I was eligible to go into the Army. I passed the exams but received a draft deferment, because of my essential work in the shipyards.

Getting back to the time I was neglecting to treat my cough and cold, I contracted bronchitis, which developed into pneumonia and became deathly sick. Providentially, my brother Joe happened to come by for a visit to my apartment in the Flatbush section of Brooklyn. I was married to Millie at the time and my brother Joe was not aware that I was home in bed seriously ill. I had little Jack who was a most beautiful and brilliant child then. Brother Joe and his girlfriend, who later became his wife, took one look at me and immediately called for an ambulance. I was rushed to Kings County Hospital somewhere in Brooklyn. This actually saved my life, for I would have died shortly. God did not want me yet and so I survived.

The first day after the very night that I was taken to the hospital, my mother and brother-in-law, Sam Cohen, came to visit me. After a few words, I asked my mother to bring a certain Rabbi who used to visit us when I was living at home.

He was a very sweet and gentle soul and I felt that his presence would comfort me. Upon hearing this, I heard my dear mother telling brother-in-law Sam that I was dying, while she was weeping, "My Hymie is dying." After a while a nurse came in and said it was time for everyone to leave. I saw my mother walking out of the room still weeping.

The next day, my mother visited me bringing a jar of chicken, vegetable soup, fresh made. My mother was an excellent cook, but I had no appetite whatsoever. Even the thought of food nauseated me, but out of respect for my mother and not wanting to hurt her feelings, I ate most of it. I did not even taste the hospital food. The next day my mother visited me again, bringing some more fresh made vegetable chicken soup. I wasn't the least bit hungry, but not to hurt her feelings I ate it once more. I still did not touch the hospital food. Then I told my mother, that instead of bringing me food, to bring me a big bag full of oranges, lemons and grapefruit, which she did the very next day. I told her that every time she would come to bring me more of the same. I believe she visited me every other day, bringing me some more oranges, grapefruit and lemons. I asked the nurses on duty round-the-clock, to find the biggest glass and fill it with a fresh squeezed cocktail of mixed juices. I drank these juices daily for twenty-seven days, taking no food whatsoever. In other words, I fasted for twenty-seven consecutive days on acid fruit juice and a little water, nothing else. For twenty-seven days I did not have so much as a glass of milk, a cup of coffee, or a single slice of toast. Of course I did not tell my mother or the nurses. I simply gave my hospital meals to others in the large rooms of four beds and patients.

From my study and practice of the Bernarr McFadden lessons on natural methods for building health and curing disease, I learned about the efficacious effects of fasting and vegetarianism, as well as all of the natural aids to health and curing of diseases of every kind. Bernarr McFadden owned and ran a Physical Culture Hotel in Dansville, New York where his system was practiced. He literally cured thousands of people of various diseases which the medical profession said were either incurable, or could be helped on a limited basis. Even many thousands of people without a disease would go there to improve their health. In my practice of his methods in the past, I had proved to my self that he was totally right. He was a true Guru of Health.

Long before Dr. Linus Pauling came out with his theories about high doses of Vitamin C for the cure of colds of any severity, I had been practicing it for years. Bernarr McFadden had been teaching the same for fifty years previously. Among true followers of Physical Culture, Dr. Linus Pauling's theories and practice were common knowledge. Even I knew more that the good doctor did. Let me prove my point.

Going back to the hospital, where I fasted for twenty-seven days on nothing but fresh squeezed acid fruit juice and water, the time came for me to be discharged from the hospital. It was customary for every patient being discharged to receive a complete check-up including X-rays.

I was in the office of the doctor who examined me and had the X-rays of my lungs with her, a female physician of course. She looked at me with dismay. She could not understand why my lungs showing up in the X-rays were perfectly clear. She put the X-rays on the light screen once more, commenting, "I can't understand it, after a severe bout with pleurisy and pneumonia, lesions would be left on the lungs, with unmistakable clarity, visible on X-rays." She said, "Your lungs are as clear, pure and healthy as those of a new-born healthy baby. I can't understand it."

It was then that I told her that I had fasted for twenty-seven days, on acid fruit juice and water. I explained that I had learned all about the fasting method with large doses of natural vitamin C for the cure of colds, bronchitis and pneumonia. In fact fasting, properly conducted and completed will cure all cases of acute disease and most cases of chronic illness. Bernarr McFadden, proved this to be true for fifty years. But the medical profession sneered at McFadden and his natural methods of cure. Why? Because it hurt their pocket books and those of the pharmaceutical industry. I proved to myself that I was a Master of Physical Culture in both building health and curing disease, superior in some ways to the medical profession.

Fasting, rightly conducted and completed is nearest a panacea for curing all disease of any other single method of cure. Bernarr McFadden proved it thousands of times as well as a number of Naturopathic physicians. I proved it to myself in curing an occasional cold and a severe case of pleurisy and pneumonia.

The Arab physicians of the Tenth and Eleventh centuries, prescribed three week fasts for the cure of all diseases, in conjunction with their prescribed herbal medicines and right diet thereafter.

Athenaneous, a Greek physician in Rome (A.D. 50) said, "Fasting cures diseases, makes the mind clearer and purer and raises man to the throne of God." He perceived the physiological mental and spiritual benefits of Fasting.

I noticed that in my twenty-seven day fast, I not only cured myself of disease, but that my mind did become clearer and purer and I attained a deep sense of peace and emotional equilibrium.

Again when I was nineteen and twenty I had the idea of going to India and find a Spiritual Master or true Guru, stay and live with him or nearby, and serve him like a servant, and would be his disciple at the same time for about seven to ten years. I had read about these Yogi Masters living in or near the Himalayan Mountains. I was saying to myself, "First the pathway to God, then the affairs of the world.' The average individual thinks, in effect, first the affairs of the world then the pathway to God if at all. If he is not spiritually inclined, then finding God does not even enter into the picture.

For the materialistic individual not concerned about God, then his way is best for him to attain whatever goals or success he envisions. He is relatively happy going his way, with periods of happiness and unhappiness, pain and pleasure rich or poor, in health or in sickness, prominent or unknown, success or failure, and all of the other dualities that the worldly life entails and are experienced.

As for me, I did not function well in the mundane world, so for me the spiritual path was more practical. I wanted to go to India and find a true Guru or Spiritual Master, realizing that if I could become centered in God and attain Cosmic Consciousness, I could then enter into the mundane world with the stature of symmetry and could handle the affairs of the world with masterful ease. I was fully aware that to attain Cosmic Consciousness and become one with God was the ultimate goal of life itself. Cosmic Consciousness is not a theory, a figment of the imagination, some hallucination or poetic dream.

It is as real as the Empire State Building or the Rock of Gibraltar. Even more so, because a nuclear bomb dropped on each one would turn them both into clouds of dust. But the Cosmic Consciousness is eternal and nothing can change it or put so much as a scratch on it. It is the mind of God Himself.

Einstein once said, "I would like to know the thoughts of God...the rest are just details." Not even Einstein can know and understand the mind of God because it takes a Cosmic Mind to know and understand the Cosmic Mind of God.

Only by attaining the Cosmic Consciousness can one know God, experience Him and become one with God. The term Yoga means "union with God", which is the goal of the Hindu philosophy and religion. It is also the goal of Buddhism.

The two processes of evolution and involution, which means the development of consciousness, are meant to attain the Cosmic Consciousness, while we are still in the human form.

Technically speaking, the process of evolution ends when man attains the human body, evolved from animal forms. Then starts the process of involution, which is the development of consciousness. For example, compare an aborigine in Australia with an Albert Einstein, dozens of incarnations lie in-between the two.

Let me state unequivocally, that the desire to attain Cosmic Consciousness and become one with God, is the greatest of all desires. When one attains oneness with God, then all right desires are satisfied completely forever and you enjoy everything but without attachment.

Spinoza implied the same philosophy when he said, "The union of the soul with God is our 'second birth' and therein consists man's immortality and freedom. "He was known as the God-intoxicated philosopher. "I am the God-intoxicated thinker."

Dante implies the same when he says in Canto IV, "Instruct me Master and most noble sir, better to understand the perfect creed that conquers every error."

Cosmic Consciousness cannot be investigated by the methods of scientific research, since such research depends ultimately on sense perception. Cosmic Consciousness is beyond the grasp of the senses; this is the certainty of certainties.

It has been the general experience of all true Gurus or Spiritual Masters, that first they had to attain an intellectual knowledge of God and the spiritual path leading to Him. Then had to develop a love for God and a longing to attain Him equal to one in the desert thirsty for water. Was I not that little child in the crib thirsting for the Water of Truth and not the drinking water?

In the past, but of course no longer true today the greatest puzzle to me in the Universe was I to myself. How could I possess the wisdom of the ages on the one hand and be so ignorant, weak, fearful and inept in the mundane world at the same time. I simply lived in two worlds that I could not harmonize. In the world of Philosophy, Literature and Scripture I was like a fish in water. In the mundane world I was a fish out of water, gasping and wanting to return to its native element. For years, I feared that if I got too involved in worldly life I would lose sight of my spiritual desire and goal. This fear of entanglement was all wrong and contributed to my failing to fulfill all of my duties and responsibilities as a father and husband. The memories are painful even to this day. I thought to myself, why can't I be like everyone else, pursuing and attaining material goals and not having spiritual ideals, which brought pain in not realizing them. In effect, I was a failure in both the spiritual world and the mundane world, not having attained success in either world. I was in limbo. My head and heart were aimed at Heaven, but I didn't have my feet on the ground. I know now, of course, that one must succeed and harmonize both worlds in order to be a true and all-round success.

Even the most successful businessman, financier or professional person, is not a true well-balanced success unless he is firmly grounded in spirituality. Because if one, by the vicissitudes of life, loses what he (or she) has attained or gained, one becomes despondent, miserable, unhappy - can lose his mental stability and emotional equilibrium, lose all of his money or health or both and divorce often steps into the picture if married. If grounded in spirituality, one becomes detached from all material and worldly things and so can handle all vicissitudes without pain or fear or worry and maintain one's poise and peace of mind. There are many factors that enter into the equation for success, but without spirituality one's life remains on shaky ground and a weak foundation at best.

The worldly life is like a dangerous ocean to cross and reach the safe shores of God and His Kingdom. There are seven passions or seven wild horses that need reigning, namely, lust, greed, anger, hatred, false pride, envy and attachment to material things and pleasures of every description with it's endless round of pleasure and pain. These attachments may also be called habits or addictions that keep us in bondage to the world. Thus the materialistic individual believes that the world and worldly life is the only reality and that God, the spiritual life and spiritual world seem unreal, vague and hard-to-grasp.

Thus the materialistic individual takes the world seriously and God lightly, whereas, the spiritually oriented individual takes God seriously and the world lightly.

The earth-life is the junction between Heaven and Hell. This is where God gives us free will, to choose which path to follow. The choice is ours and thus we make our own Heaven and our own Hell.

Let me relate a significant incident that occurred many years ago. I made a trip to California to see my son Jack and stopping in Chicago for a few days to visit my sister Vicky and her married daughter with children. Arriving in San Francisco I recall staying with Jack and his then girlfriend in their apartment for about two weeks. It was time for me to move with my suitcases elsewhere. I wound up at the YMCA where the rates were cheap by the week. My funds were running out and in order to stay on longer, I got myself a job in one of the downtown banks on lower Market Street, as a guard. I was paid by an employment agency not directly by the bank.

A Vietnamese or Cambodian individual also worked at the same bank, also as a guard. We became friendly and talked about experiences and ourselves. One day he told me of his life in Cambodia, where he was in a guerrilla army, fighting the Communists, who were cold-blooded killers of civilians and military alike. He said he was the leader of a troop of men and had killed many of the enemy himself. I knew that he wasn't boasting or exaggerating, as he was simple and straightforward. Also, he told me that he had been married and brought his wife with him to the USA about seven years previously. Then he told me that he had a friend also from Cambodia, who had come to this country and friends of his wife and himself. The saddest part of his story was that his friend had seduced his wife and urged her to divorce her husband and marry him. The rival was earning much more money than he was and could provide her with a much better lifestyle. His wife did leave him to live with this rival. He was sad and angry over this betrayal and brooded over it for some time.

Then he told me that in his anger and desire for revenge, he was thinking seriously of killing his rival. He even showed me the knife he was carrying and he told me that he was a Master of Karate by which he killed many of the enemy in Cambodia. He could have killed his rival in a split-second. After listening to him thoughtfully and intently, he asked me what I thought of the whole matter. I told him to forget his anger and desire for revenge; that there was a Law of Compensation or Retribution, which would exact the penalty for that betrayal. I said that he should be thankful that he didn't commit such a sin on someone else and that someone else did it to him instead. Therefore, the Retribution would fall on his rival's head and not on his own head.

He marveled at my sage-like advice, thanked me and said "You speak wisely like my father, my father would have liked to have met you." Then he said, "I am going to do something that I have never done before. We were both sitting on chairs during this whole conversation. He got up from his chair, got down on his knees and touched his hands and forehead to my two feet. This is the tradition in the Hindu and Buddhist religions, where people and disciples do the same to their Spiritual Master. The reason is that in doing so according to their traditions, the people lay upon the Master, the burden of their sins.

This is a form of spiritual cleansing and symbolizes the surrender of their wills to the Master. The feet, though the lowest part of the body, are considered to be the highest form of the spiritual point of view.

This Vietnamese or Cambodian fellow employee was paying his respect for me as his Spiritual Master. When he got up he smiled and looked relieved as though he had laid his evil thoughts of committing murder at my feet. His face shone as though he had been cleansed and I'm sure that he never had the same wrong thoughts again. He was 'reborn' in the same sense that the story of the bully was in my early years.

When I was about forty, I had moved back to Bridgeport, CT from Port Jervis, NY where I had worked for my two brothers, Sam and Joe. They were manufacturers of ladies garments and quite successful. Back in Bridgeport, I got a job with Sikorsky Helicopters as a sheet-metal fabricator. Since I had worked in the shipyards as a ship fitter, the work was similar. Instead of working with steel, I was now working with aluminum and titanium. I worked there for four years, when there was a general lay-off. I got a job immediately with Avco-

Lycoming as a sheet-metal fabricator, also working on steel parts and assembly work. After twelve years, there was a general lay-off. Let me say that Avco-Lycoming made the engines that went into military helicopters, since these engines were very expensive and only the government could afford them from the vast military budget. Foolishly, when I was called back, I did not return to my job, having got entangled with other pursuits to make a living. This proved to be a disaster I didn't realize that I was much better off working for a large company than working for myself. I was ignorant, weak and inept working for myself and not smart, sharp or strong. I even lacked faith in myself.

There are three kinds of faith that are most vital,

1. Faith in Nature, which ensures health.
2. Faith in God, which ensures abiding peace of mind and heart.
3. Faith in Self, which ensures success.

My Faith in Nature and God were perfect, but I lacked that strong faith in myself to be successful in the mundane world. I was like a lop-sided tire, with a big bulge on one side, making it impossible to spin right. Even the wheels on a car need periodical balancing in order to turn smoothly. I was anything but balanced thus becoming generally unhappy even miserable.

The saddest part of my life came during the years that I worked for Avco-Lycoming in Stratford, CT, which was next door to Bridgeport, CT where I lived with my dear wife Ruthie and two wonderful sons, David and Michael. During those years, Ruth developed cancer. The doctor advised an operation. My dear Ruth lived for about a year and one half after the operation. On the day that she died in the early morning hours of October 20th, it was Mike's birthday. I can never forget that morning when I told Mike that his mother had passed away; he burst into tears saying, "She had to die right on my birthday!" He broke my heart. The memory of it still breaks my heart, to this very day, to this very moment, I am writing it. If I were to live to be a thousand years old, I can never forget that day. It was indeed the most tragic event of my life.

When I realized that she was not going to survive, I would awake in the morning as the sun shone and look out the window and pray, "Dear Lord, if one of us must go, let it be me, let it be me." I said this prayer more than once. I was thinking that children need their mother more than their father. Nothing can compare with a Mother's love for her children. Besides, believing in Reincarnation and the immortality of one's soul, I thought that I would get another opportunity in my next life on earth to attain the goal that I prized the most - attaining Cosmic Consciousness and oneness with God. This idea and ideal has stayed with me for over sixty years and has never changed. If anything it has become clearer, stronger and I have become more determined to attain it in this incarnation.

Before my dear Ruth passed away, one morning sitting on the bed next to her lying down she told me, "I love you very much" and I told her that I loved her too and that I wasn't worthy of her. I told her "I am not worthy of you, you are too good for me." She answered in protest, "No, no you are good enough." We meet our loved ones again in Heaven and I will be with her once again as well as with my dearest Mother and Father and all others. God Himself is Love and Love is the only coin that gets us into Heaven.

Besides losing Ruth, my other great loss was losing the companionship of little Jackie, when Millie and I were divorced. I can honestly say that I never wanted a divorce. It never even entered my mind. But Millie insisted on it and I just went along with her wish. I could still have kept in touch with my dearest Jackie (I always called him Jackie) but I neglected to do so and I have always regretted it. The memory of it pains me whenever I think of it to this very day. But I thank God that I have him today. I love the three of you equally and dearly and nothing gives me more pleasure than thinking of you individually and collectively. No amount of success or money can ever replace my love for the three of you. God never gave any father three brighter, nicer, sweeter and more lovable boys than my three sons. It is, in fact, God Himself who comes to us in the form of our children. Are not our soul's a part of the Supreme Soul of God?

Therefore, we should welcome, love and adore, our children as God in human form. See God in each other for we are all a part of God. The bottom line is that God is the one and only Reality. The entire universe is the body of God and if our souls are a part of God, what is left? Only God! And nothing can exist without God, and nothing can exist outside of God.

The entire Creation came out of the Mind of God and He sustains it. The Creation needs God to exist, but God doesn't need the Creation to exist. He is independent and self-sustaining.

Ralph Waldo Emerson calls God or the Supreme Soul - the Over Soul.

Swami Vivekananda, a Yogi Master of the early part of this century and a disciple of the Spiritual Master Sri Ramakrshna said, "The highest Truth is this; God is present in all beings. They are His multiple forms. There is no other God to seek. He alone serves God who serves all other beings. Is Swami Vivekananda not saying in his own words that our souls are a part of God that He is the only Reality and that only God exists as the essence of everything? Only God is Real. Everything else is in the state of flux or the process of change through Evolution and Involution, and therefore has only a relative or temporary reality and is not real. Since God is permanent and never changes, therefore only He is Real, the one and only Reality.

The long journey of the soul consists in developing through Evolution, from animal consciousness to human consciousness and then transcending the limited human consciousness and attaining the unlimited Cosmic Consciousness. What we call Life is only the development of Consciousness going from the unconscious to the conscious, from unconscious God to conscious God.

We come to the realization that we are a part of God and that our souls are no different than God, qualitatively speaking, just as the water in a wave of the ocean is no different, qualitatively speaking, than the water in the entire ocean. With this analogy, we can say that every one of us is a wave in the Ocean of God Himself and that God is everything and in everybody.

Mathematically speaking, think of God as being a very large equilateral triangle. Think of us as being equilateral triangles coinciding with the very large equilateral triangle. Also, each triangle coinciding with the very large triangle is still an individual triangle. Now when one attains the Cosmic Consciousness, one's soul blends with God as the wave in the ocean blends with the ocean, but at the same time we maintain our individuality like the illustrated coinciding triangles. This is the Magic of God - the Supreme Magician. He also plays the game of hide and seek. I hide and you seek ME. It is His game and sport and we only have to play it until we win. We can only find Him when through the development of consciousness (involution), we finally attain the unlimited Cosmic Consciousness. Our limited ego-mind is the veil that hides God from us. The limited individual ego-mind of ours must be completely submerged and dissolved into the unlimited Mind of God or the Cosmic Consciousness. When the limited ego-mind disappears, then we attain the God-state, the Yoga or union with God, the goal of the Hindu religion and philosophy. It is exactly the same goal of the Buddha with his Eightfold Path to Perfection.

This world is not our true home. This planet earth is just a school of experience where we learn how to behave and mature spiritually until we are fit to live with God in His Kingdom, which is our true and Eternal Home with Eternal Life. This is not a beautiful dream, a theory or poetic imagination. It is a spiritual Reality as the Law of Gravity is a scientific reality. This is the Perfection that the Buddha preaches in his own words.

Plato implies the same Truth when he says, "If there exists a good and wise God, then there also exists a progress of mankind towards perfection." Plato is also implying that one cannot separate God from perfection because they are heads and tails of the same coin. From the Perfect God can only come what is perfect. Be God-oriented, not world-oriented.

In the beginning only God existed like a placid ocean. Then when God stirred and spoke the creative spiritual word 'Om', the ocean became full of waves and those waves became the Creation and our individual souls, all part of the ocean and not separate, like the waves in the ocean are not separate from the ocean. So the universe and everything in it springs into existence from the Om Point. God's only purpose for His Creation is to develop consciousness to its highest point. It is His Play of Consciousness and He is the Author and Director; the theater, the stage and the actors are all His Creation.[277]

As long as I can remember, I have always had a clear, spontaneous, concise and intuitive perception of Truth, expanding more and more as I grew older and older. My intellect simply expanded on the Truth that I perceived easily and intuitively. When the intellect coincides with intuition, then we know we are absolutely right on target. As my intellect expanded till the age of nineteen and twenty, I knew then that the highest Truth was the principle of Cosmic Consciousness, that it was absolutely possible to attain and totally desirable to experience. That's why I wanted to go to India, meet a true Yogi or Spiritual Master and attain the goal. I had read a book titled, 'Cosmic Consciousness' by an author named Bucke, which may be still available in Main Libraries. As a member of the Rosicrucian Order, this ideal was expressed innumerable times as part of their philosophical concepts. The term became a part of my consciousness like the world was round and not flat. My intuitive sense and intellect combined said it was the certainty of certainties and no one could dispute it with me. One's spiritual certainly cannot be challenged by anyone. The materialistic world may dispute it, but then the Mass Mind is spiritually ignorant. One may be religious and ignorant but one cannot be spiritual and ignorant.

This became my goal in life at the comparatively early age of nineteen-twenty, and has never changed since, in over sixty years. An authoritative and ultimate ideal lasts a lifetime and never needs renewing. The hardest thing in life to do is to attain ones own ideal. Endless years of struggle, anguish and failure accompany the man with a high ideal.

An ultimate ideal last a lifetime and never needs renewing to repeat. It is indestructible. But we must translate our ideals into the vernacular of life and attain the goal.

The ancient sages of India called the soul simply the Self. The doctrine of Self-Realization says that God dwells within the body as the Self and that the body is the temple of God-the Self, The Lord, they said, is indeed the dweller in the body as the Self.

Lord Krishna says in the Bhagavad Gita "Steady in the Self, being freed of all material contamination, the Yogi achieves the highest perfection state of happiness in touch with the Supreme Consciousness." (Cosmic Consciousness)

The spiritual aspirant, following the path of meditation turns within. The inward directed movement finally, with practice, finally merges the mind into the inner Self and melts into the plane of Cosmic Consciousness, which is the goal of meditation. Meditation and Mantra repetition are meant to attain this highest

---

[277] Herman Melville's "Moby Dick" in "Loomings".

consciousness. It removes the veil that separates our limited consciousness from the Cosmic Consciousness or God- Consciousness.

A Spiritual Master gives the Mantra to us. Right understanding, meditation on the Self within is the way to find God and unite with Him. This ultimate goal cannot be realized independently of Creation. The individual soul undergoes the experience as an individualized ego and limited mind through many incarnations until it matures spiritually and is ready to attain the Cosmic Consciousness and is freed from the recurring births and deaths or Buddha's Wheel of Rebirth or Reincarnation. Then, when one attains the highest goal, the Cosmic Consciousness, one no longer is compelled to re-incarnate on earth and lives eternally in the Kingdom of God with God. Eternal Life consciously in God's Kingdom is an absolute reality. Then the dear, wise purposes of God are fulfilled in His Play of Consciousness. For such a one, the Play is over!

Meditation may be defined as the path that the spiritual aspirant takes trying to get beyond the limitations of the mind to the inner Self. With the practice of meditation, the aspirant will turn more intensely to God.

Past wrong thoughts, wrong words and wrong actions have left 'seeds' or impressions on the mind and one cannot achieve emancipation of the soul until these impressions or 'seeds' are burned in the fires of wisdom and meditation. Meditation is our link with God and the means to find God within-the Self. It is wise to seek God before our 'earth-visa' expires. Otherwise, we have to return to earth (reincarnate) and suffer all over again to a greater or lesser extent and experience all the dualities again, pain and pleasure, happiness and misery, health or sickness, rich and poor etc.

Milk poured into water readily mixes with it. Similarly, the waters of ignorance, weakness and illusion quickly dilute the milk from an ordinary person's mind. The Yogi, the man of spiritual self-discipline churns the milk of his mind into the butter-state of divine stability and is able to float serenely on the waters of worldly life.

He is as the maxim says, "To be in the world, but not of it." This is the hardest thing to do. But when one attains Cosmic Consciousness, it becomes like child's-play. This Cosmic Consciousness makes of this world a little playroom, of mortal possessions a box of toys, of the human race a handful of tin-soldiers, and of you, owner of the nursery, dispenser of the toys, Commander of the host.

In the conscious state of Cosmic Consciousness, one becomes a Man-God; man and God at one and the same time. We rise up from being a limited and ordinary human being into Consciousness, Knowledge and Power and being a Son of God.

All who have reached the highest Consciousness, the Cosmic Consciousness are no longer ordinary human beings, but humanly divine and divinely human. Call him a 'Cosmic Man' or a 'Cosmocrat'. He is the real Superman and not the comic-book character. So, never underestimate this ideal of attaining Cosmic

Consciousness, which is the highest and greatest ideal possible to attain on this Planet Earth.

God makes His devotees like Himself in order to bring others back to God.

The Spiritual Master or Cosmic Conscious Man has only one desire, to teach and raise others to his own level. It is God's Plan and desire to bring all human beings, His children back to Himself in God's Kingdom with Eternal Life. The ultimate goal is attained individually. Everything fails on earth but the Cosmic or God-Consciousness where there is never any failure, disappointment, ignorance, weakness, pain, sin or misery. There life becomes a triumphal march to the citadel and royal domain called God-Consciousness.

I have used the word 'intuition' many times, and it is essential that I explain what 'intuition' is and how important it can become in one's life.

Intuition is the voice of the past. That is, what we get by reason in this life, we get by intuition in the next life. Reason can grasp only the cause-effect principle that pertains to the phenomenal world; higher than reason is intuition, knowledge derived immediately and spontaneously from the soul-not from the fallible agency of the senses or reason. Intuition is uncommonly known as the 'sixth-sense', by which one apprehends knowledge distilled from the multitude of experiences, gathered from past lives and which become part and parcel of the intuitive make-up of one's active consciousness in this life. The details of one's past lives are eliminated; just the distilled essence remains as intuitive wisdom.

In other words, intuition is the digested experience of the past lives, or we may say it is the compressed and consolidated wisdom distilled from the past. Intuition is actually a 'voice of the soul'. The soul 'speaking' to the body is called instinct (animal instinct). The soul 'speaking' to the heart is called intuition. The soul 'speaking' to the mind is called inspiration and last comes revelation, which is the soul immersed in the Divine Light and in which Divine Knowledge and understanding comes from God Himself. This revelation is also known as 'illumination of the soul'. Thus, it can be said truly that intuition is superior to reason. We still need reason for our current experiment. In other words, intuition tells us what to do, reason tells us how to do it.

In my dreams as a boy of seven in the bottom pit of Hell facing the Devil, how did I know that God would protect me and keep me safe and had nothing to fear from Satan? Feeling the very Spirit of God surging through my body and making me absolutely fearless and challenging the Devil to do his worst, how did that come about? It was my intuitive heart telling me that God was my Friend and that no one or nothing could hurt me. Where did my supreme faith in God come from without thought or learning? How could I read Scriptures at ten-eleven with ease and understand everything if I hadn't learned these things in a former incarnation?

The teachings in the Bible were to me as clear as crystal, never doubting its Truth for a single solitary second. I even felt that Jesus was not some distant historical character, but more like a friend or relation of my family. I must have had an incarnation at the time of Jesus, otherwise we would not have been, both of us, born in caves and wrestled with Satan and emerging victorious. That is not a coincidence. I even think like Jesus - we are on the same wavelength.

At nineteen-twenty I had, intellectually and intuitively, the highest spiritual wisdom of the ages and knew what the highest and ultimate goal of life was - to attain God and become one with Him. Nothing mattered but supreme faith in and love for God - the one and only Reality. Only God is Real. Even our worldly life is an awake-dream like we have a dream-dream. We only become really awake when we awaken in the Cosmic Consciousness, when the ultimate Truths of the Infinite are revealed to us in a glorious array. *You ain't seen nothing yet!*

XIV.

"Not from the stars do I my judgment pluck;
And yet methinks I have astronomy,
But not to tell of good or evil luck,
Of plagues, of dearths, or seasons' quality;
Nor can I fortune to brief minutes tell,
Pointing to each his thunder, rain and wind,
Or say with princes if it shall go well,
By oft predict that I in heaven find:
But from thine eyes my knowledge I derive,
And, constant stars, in them I read such art
As truth and beauty shall together thrive,
If from thyself to store thou wouldst convert;
Or else of thee this I prognosticate:
Thy end is truth's and beauty's doom and date."
Shakespeare

Photo of Jack at 62 by A.T. Conway[278]

Prospero's Cell, Shakespeare's "Tempest"

Our revels now are ended. These our actors,
As I foretold you, were all spirits and
Are melted into air, into thin air:
And, like the baseless fabric of this vision,
The solemn temples, the great globe itself,
Ye all which it inherit, shall dissolve
And, like this insubstantial pageant faded,
Leave not a rack behind. We are such stuff
As dreams are made on, and our little life
Is rounded with a sleep.

---

[278] He might very well pass for 39 in the dusk with the light behind him! Stolen from Sir W.S. Gilbert.

Zero Point Energy, Star Gates & Warp Drive

Notes on "The Casimir Effect
Physical Manifestations of Zero Point Energy"
K A Milton, World Scientific (2001)
Under Construction

"When in doubt, integrate out."[279]

By Jack Sarfatti

---

[279] "Roger Rabbit Goes To College" ;-)

Abstract

This paper contains original important discoveries not found anywhere else. It addresses and answers the following questions:

Why classical curved spacetime at all? (Sakharov, 1967)

What is the "dark energy" that is most of the mass of the universe invisible to electromagnetic detectors yet detectable gravitationally?

Why is the universe's expansion rate speeding up rather than slowing down?

What is Kip Thorne's "exotic matter" needed to keep traversable wormholes open?

Is vacuum propulsion of unconventional flying objects[280] possible in principle?[281]

In the course of answering these questions I derive the detailed nature of the coherence mechanism of the zero point energy vacuum fluctuations. What we have is an electrically neutral spin 0 Bose-Einstein condensate[282] of mainly virtual electron-positron pairs. I derive explicit formulas for the parameters of the effective potential of this condensed phase of quantum vacuum without which classical gravity could not even come into being. These formulas show implicitly how the quantum phase transition leading to inflation happens. They also suggest how to control the inflation process on a small scale. The implications of such a development are, of course, profound.

---

[280] Book by Paul Hill a USG aeronautical engineer in 40's and 50's is reliable.
[281] In the sense of Alcubierre's warp drive on a free float timelike geodesic without harmful g-forces.
[282] Classical curved spacetime is supported by a giant coherent quantum wave (or qubit field) inside the physical vacuum.

The first obvious fact from this book by K. A. Milton is that the widespread claim by the New Age "Cargo Cult"[283] Alternative Energy and UFO Disclosure Movement that the theories of Hal Puthoff mainly, also Bernie Haisch and Alfonso Rueda secondarily, are some kind of panacea breakthrough already here for world energy problems is "not even wrong"[284] without any scientific foundation. Casimir force effects are very tiny. Their most immediate applications would be to tiny nanometer scale machines. The UFO explanation by Eric Davis at MUFON 2001 is completely bogus, although the alleged phenomena cited in his paper may not be.[285]

Connection of zero point energy to intermolecular Van der Waals forces.

The interaction Hamiltonian between two electric dipoles is

$$H_{electric-dipole-dipole} = -\vec{d}_1 \cdot \vec{E}_{12} = -\vec{d}_2 \cdot \vec{E}_{21} = \frac{\vec{d}_1 \cdot \vec{d}_2 r^2 - 3\vec{d}_1 \cdot \vec{r}\vec{d}_2 \cdot \vec{r}}{r^5} \quad (1.1)$$

For a paraelectric randomly oriented ensemble of dipoles the quantum expectation value of the coupling energy from first order time independent perturbation theory vanishes.

$$\left\langle H_{electric-dipole-dipole} \right\rangle = 0 \quad (1.2)$$

In second order perturbation theory, the effective *static* dipole-dipole potential energy is

$$V_{eff}^{(2)} = \sum_{n \neq 0} \frac{\langle 0 | H_{electric-dipole-dipole} | n \rangle \langle n | H_{electric-dipole-dipole} | 0 \rangle}{E_0 - E_n} \square \frac{1}{r^6}$$

$$(1.3)$$

---

[283] Richard Feynman's famous talk at Cal Tech on "Cargo Cult Pseudoscience".
[284] Wolfgang Pauli's nasty comment on bad physics.
[285] http://198.63.56.18/pdf/davis_mufon2001.pdf

Where $|0\rangle$ is the "vacuum". Including time delay retardation gives $\sim r^{-7}$ at larger distances between the dipoles. The polarizability α is defined by

$$\vec{d} = \alpha \vec{E} \tag{1.4}$$

At absolute zero temperature

$$V_{eff}^{(2)} \square \frac{\alpha_1 \alpha_2}{r^6} \frac{\hbar c}{r} \tag{1.5}$$

The general mainstream zero point energy computational situation is murky because of the ambiguities of renormalization of subtracting two infinities to get a finite number. True, the procedure empirically works to fantastic accuracy in quantum electrodynamics, but it is one of the mysteries why it does.[286] The Green's function method seems to require the outgoing far field radiation boundary condition that appear to exclude nonradiating near fields? The Casimir force for parallel plates is attractive, but it is repulsive for a conducting neutral sphere (Timothy Boyer). Attempts to compute zero point energy effects of confined strong gluons in the microscopic hadron bag model are inconclusive. Applications to extra Kaluza-Klein space dimensions are also in a sorry state. When it comes to cosmology,

"Significant issues arise when we consider gravitation, because the absolute scale of energy presumably is now meaningful as the source of gravity. In particular, one might think that the cosmological constant would have its origins in quantum fluctuations of the gravitational and other fields, yet naïve estimates give far too large a value." p. 201 (Milton).

I solve this problem in this paper. I should give some credit to Giovanni Modanese whose work with real superconductors, not the virtual one in the superfluid quantum vacuum, gave me one of the key ideas for this

---

[286] See Feynman's popular books especially the one on "Quantum Electrodynamics" with all the little arrows.

solution. Another key idea was Hagen Kleinert's [287]"solid state" formalism for general relativity in terms of a "world crystal lattice" with curvature and torsion as string topological defects of disclination and dislocation respectively. I also should give credit to Hal Puthoff and Bernie Haisch whose suggestions on the zero point origin of inertia and gravity seemed so silly and wrong headed to me that I decided to solve the Sakharov problem that motivated them correctly. Basically none of the Pundits, not just Hal and Bernie, had the correct qualitative picture of the actual quantum vacuum although fragments like QCD gluon condensates were in the air pointing in the right direction. No one, until me, realized how to use Bohm's quantum realism to derive Einstein's 1915 geometrodynamics from the phase modulation of the vacuum virtual Bose-Einstein condensate with the quintessent $\Lambda(x)$ field from its amplitude modulation.

Quintessence as finite renormalization?

The factor $\hbar c/r$ is a quantum zero point fluctuation term. I suggest for Popper falsification, the completely new original empirical rule for $L_p^2 \equiv \hbar G/c^3 \,\square\, 10^{-66}\, cm^2$

$$\frac{\hbar c}{r} \to \frac{\hbar c L_p^2 \Lambda(\vec{r},t)}{r} = \frac{\hbar c}{r}\left(1 - L_p^3 \left|\Psi(\vec{r},t)\right|^2\right) \tag{2.1}$$

where $\Lambda(\vec{r},t)$ is the local quintessent field and $\Psi(\vec{r},t)$ is the local order parameter of the physical vacuum describing the Bose-Einstein condensate of virtual off mass shell Goldstone "tachyons" with

$$M^2 < 0, \beta > 0 \tag{2.2}$$

in the infrared

---

[287] http://www.physik.fu-berlin.de/~kleinert/

$$\left|\vec{k}\right| << Mc/\hbar \tag{2.3}$$

where

$$\omega^2 \neq (kc)^2 - \left|Mc^2/\hbar\right|^2 \tag{2.4}$$

The local macroscopic quantum phase coherent vacuum order parameter obeys the spacetime + gauge covariant Landau-Ginzburg Bit From It equation

$$D^\mu D_\mu \Psi + \left(\frac{Mc}{\hbar}\right)^2 \Psi + \frac{\beta}{\hbar^2}|\Psi|^2 \Psi = 0 \tag{2.5}$$

where

$$\Psi(x) = |\Psi(x)| e^{i\Theta(x)}$$

$$|\Psi(x)| \equiv \sqrt{\frac{1}{L_p^3}\left(1 - L_p^2 \Lambda(x)\right)}$$

$$\underset{\Lambda \to 0}{Lim} |\Psi(x)| \to \sqrt{\frac{1}{L_p^3}} \neq 0$$

$$\underset{\Lambda \to 1/L_p^2}{Lim} |\Psi(x)| \to 0 \tag{2.6}$$

Note that the completely incoherent locally random hugely would be antigravitating vacuum[288] of Flat World indeed has $\Lambda \approx L_p^{-2} \approx 10^{66}\, cm^{-2}$.

---

[288] This is a counter-factual definite statement used in quantum theory. See Roger Penrose's popular books. Example, "I would if I could, but I am not able." (Pirates of Penzance) Something that might have happened, but didn't, would have been definite if it did. This is no joke, e.g. Elitzur-Vaidman land mine tester p. 268 Penrose, "Shadows of the Mind".

Possibility 1: singularities are real.

$$-\infty \le L_p^2 \Lambda(x) \le 1 \tag{2.7}$$

If this is how the universe works then we have a singularity when

$$\lim_{\Lambda \to -\infty} |\Psi| \to \sqrt{\frac{|\Lambda|}{L_p}} \to +\infty \tag{2.8}$$

Possibility 2: singularities are an illusion.

$$-1 \le L_p^2 \Lambda(x) \le 1 \tag{2.9}$$

This means that if L_p is an absolute universal cut off for all continuum based local quantum field theories

$$|\Psi|_{\max} = |\Psi|_{\Lambda=-1/l_p^2} = \sqrt{\frac{1}{L_p^3}\left(1+\frac{L_p^2}{L_p^2}\right)} = \sqrt{2}\,|\Psi|_{\Lambda=0} \tag{2.10}$$

That is, the maximal possible Goldstone virtual tachyon Bose-Einstein density is only twice the corresponding density at the Einstein limit of 1915 where $\Lambda = 0$. It's beginning to look like all the infinities of renormalizable quantum field theories can be made finite using the local quintessent field $\Lambda(x)$. I make this as a conjecture.

"It From Bit" + Bit From It = "Universe as a Self-Excited Circuit" [289]

Einstein's classical curved spacetime metric field obeys the deBroglie-Bohm-Josephson gauge covariant "phase lock" between "It" particle and "qubit" pilot wave[290]

---

[289] John Archibald Wheeler

$$g_{\mu\nu}(x) = \frac{1}{2}L_p^2\{\tilde{D}_\mu,\tilde{D}_\nu\}\Theta(x) = \frac{1}{2}L_p^2\left[\tilde{D}_\mu\tilde{D}_\nu + \tilde{D}_\nu\tilde{D}_\mu\right]\Theta(x)$$

(3.1)

where

$$\tilde{D}_\mu\Theta \equiv \left(\frac{\partial}{\partial x^\mu} - \frac{e}{\hbar c}A_\mu\right)\Theta$$

(3.2)

$$\vec{D}\Psi \equiv \left(-i\vec{\nabla} - \frac{e}{\hbar c}\vec{A}\right)\Psi \;\&\; D_0\Psi = \left(\frac{i}{c}\frac{\partial}{\partial t} - \frac{e}{\hbar c}A_0\right)\Psi$$

$$D_\mu \approx -i_{(\mu)}\frac{\partial}{\partial x^\mu} - \frac{e}{\hbar c}A_\mu + \Gamma...$$

$$i_{(1)} = i_{(2)} = i_{(3)} = i, i_{(4)} = -i$$

(3.3)

There is a perennial tension between classical Einstein relativity and quantum theory. For example
- Locality of classical relativity vs. quantum nonlocality
- Non-renormalizability of quantum gravity

Here we have the requirement of classical relativity's local covariance in tension with the Hermitian operators in qubit space requiring real eigenvalues for quantum observables like the momentum and energy of a real particle[291] (quantum). This requires using the signature operator $i_{(\mu)}$ to maintain the Minkowski local light cone structure in $\theta \approx p^\mu x_\mu/\hbar = (\vec{p}\cdot\vec{x} - Et/c)/\hbar$ that demands $\vec{p} \to \hbar\vec{\nabla}/i, E \to i\hbar\partial/\partial t$,

---

[290] This is the elastic-plastic strain tensor of Hagen Kleinert's 4-Dim "World Crystal Lattice" of scale L_p for the unit cell. 1-Dim string defects of disclination are the curvature of gravity. 1-Dim string defects of dislocation are the torsion gaps breaking "graviton" closed strings in Curve World into "gauge boson" open strings in local Flat World and vice versa.

[291] On the mass shell, i.e. pole of Feynman propagator in complex energy plane of underlying Flat World.

otherwise we would be in the Euclidean spacetime from a Wick rotation[292]. There is no summation convention between the signature operator and the ordinary partial derivative in (3.3) and the Γ term is context dependent on the rank of the tensor operand. For example, for the scalar field $\Psi(x)$ there is no Γ term. However.

$$D_v D_\mu \Psi = i_{(v)} \frac{\partial}{\partial x^v} D_\mu \Psi - \frac{e}{\hbar c} A_v D_\mu \Psi - \Gamma^\sigma_{\mu v} D_\sigma \Psi$$

(3.4)

The fully covariant D'Alembertian back-action BIT FROM IT[293] "wave propagation" term is then

$$D^\mu D_\mu \Psi = i_{(\mu)} g^\mu_\sigma \frac{\partial}{\partial x^\sigma} D_\mu \Psi - \frac{e}{\hbar c} g^\mu_\sigma A_\sigma D_\mu \Psi - \Gamma^{\mu\sigma}_\mu D_\sigma \Psi$$

$$= i_{(\mu)} g^\mu_\sigma \frac{\partial}{\partial x^\sigma} \left(-i_{(\mu)} \frac{\partial}{\partial x^\mu} - \frac{e}{\hbar c} A_\mu \right) \Psi - \frac{e}{\hbar c} g^\mu_\sigma A_\sigma \left(-i_{(\mu)} \frac{\partial}{\partial x^\mu} - \frac{e}{\hbar c} A_\mu \right) \Psi$$

$$-\Gamma^{\mu\sigma}_\mu \left(-i_{(\sigma)} \frac{\partial}{\partial x^\sigma} - \frac{e}{\hbar c} A_\sigma \right) \Psi$$

(3.5)

Note the virtual condensate-virtual photon couplings. Every $i_{(\mu)} \partial/\partial x^\mu$ operator is a virtual Bose-Einstein interaction part. When multiplied with a A_v it's an interaction with virtual photons, when multiplied by a Γ it's an interaction with the classical geometrodynamic connection field that is the smooth collective emergent Curve World[294] mode from a spontaneous broken Goldstone symmetry in Flat World. This is the solution to Andre Sakharov's problem of 1967. One never need directly quantize the

---

[292] Stephen Hawking makes frequent use of this formal trick in quantum cosmology.
[293] This is what John Archibald Wheeler left out of his "IT FROM BIT", "LAW WITHOUT LAW" Demiurge Platonic Vision of the "UNIVERSE AS A SELF-EXCITED CIRCUIT" (illustrated in Escher's "Drawing Hands").
[294] I will *try* to use "World" for 4-Dim spacetime and "Land" for 3-Dim spacelike slice of spacetime.

gravitational field! It is a misconception to try to do so. That's why quantum gravity is not renormalizable in the global Flat World sense. It is not supposed to be! Hal Puthoff's attempt to solve this problem is "too cheap".[295]

Furthermore, the Curve World Levi-Civita connection for parallel transport of tensors along world lines in terms of the gauge force covariant Flat World derivatives is

$$\Gamma_{\mu\nu\sigma} \equiv \frac{1}{2}\left(\tilde{D}_\nu g_{\mu\sigma} + \tilde{D}_\sigma g_{\nu\mu} - \tilde{D}_\mu g_{\nu\sigma}\right)$$

(3.6)

The gapless (massless) Goldstone-like bosons[296] are from the small oscillations in the Θ quantum phase field around the *minimum* of the effective potential $V(\Psi^*, \Psi)$ of the virtual infrared tachyonic physical vacuum superfluid. The massive Higgs-like bosons with $m^2 > 0$ are from the small oscillations of the amplitude $|\Psi|$ again at the *minimum* of the effective potential $V(\Psi^*, \Psi)$ where[297]

$$V(\Psi^*, \Psi) \equiv Mc^2 \Psi^* \Psi + \frac{\beta}{M}(\Psi^* \Psi)^2$$

$$\frac{M}{\hbar^2}\frac{\delta}{\delta\Psi^*}V(\Psi^*, \Psi) = \left(\frac{Mc}{\hbar}\right)^2 \Psi + \frac{\beta}{\hbar^2}(\Psi^* \Psi)\Psi$$

$$D^\mu D_\mu \Psi + \frac{M}{\hbar^2}\frac{\delta}{\delta\Psi^*}V(\Psi^*, \Psi) = 0$$

(3.7)

The limit of Einstein's 1915 theory of gravity is when $\Lambda \to 0$. This requires a large quantum vacuum tachyonic Bose-Einstein condensate

---

[295] What Albert Einstein allegedly initially told David Bohm in 1951 on seeing his pilot wave theory.

[296] In this new original context I have here created.

[297] $m_p = \sqrt{\hbar c / G} = 2.18 \times 10^{-5}\, gm$

$|\Psi| \rightarrow \sqrt{1/L_p^3}$ for the more stable lower energy density physical vacuum that permits classical curved spacetime $g_{\mu\nu}(x)$ to come into being from the phase field $\Theta(x)$ of the new coherent order that is locally nonrandom and smooth.[298] This is "Curve World". The locally random false vacuum "Flat World"[299] is the Haisch-Puthoff-Rueda theory in which $\Psi = 0$ and $1/Lp^2 = 10^{66}$cm-2. The quantum vacuum corrected local "IT FROM BIT"[300] Einstein field equation is generally

$$G_{\mu\nu}(x) + \Lambda(x) g_{\mu\nu}(x) = -8\pi \frac{G}{c^4} T_{\mu\nu}(x)$$

(3.8)

In which $\Lambda(x) > 0$ is Kip Thorne's universally antigravitating "exotic matter" needed to accelerate the expansion rate of the universe, keep traversable wormhole "Star Gates" open and keep timelike geodesic free float "warp drives" working. The other case of $\Lambda(x) < 0$ is universally gravitating "dark energy" that is the "missing mass" of the universe. These important empirical results come from the covariant equation of state of the quantum vacuum energy, which is

$$\rho(qmvac)c^2 + p(qmvac) = 0$$

(3.9)

where $\rho(qmvac)$ is the effective "zero point" mass density of the quantum vacuum, and $p(qmvac)$ is the effective "zeropoint" quantum pressure of the vacuum. In fact

$$t_{00}(qmvac) \equiv \rho(qmvac)c^2 = \frac{\hbar c}{L_p^2} \Lambda(x)$$

(3.10)

---

[298] This was the problem posed by Andre Sakharov in 1967, but not solved properly until now by me.
[299] "Flat, stale and unprofitable" Hamlet, Shakespeare
[300] Because of macro-quantum Ψ dependence via quintessent Λ.

The active gravity mass-energy density that, for example pumps cosmic inflation in Einstein's theory, is $\rho(qmvac)c^2 + 3p(qmvac)$. The factor of 3 is crucial. When $\Lambda(x) > 0$, then $\rho(qmvac)c^2 + 3p(qmvac) < 0$, i.e. the active gravity mass-energy density is negative which means universal antigravity, i.e. Kip Thorne's exotic matter. Similarly, $\Lambda(x) < 0$, i.e. the active gravity mass-energy density is positive which means gravitating "dark energy" that is invisible electromagnetically, but not gravitationally because it is a nonclassical phase of the physical vacuum.

Local Conservation of Stress-Energy Density Currents

When Λ is constant, and there is zero torsion[301], and metricity[302] the two Bianchi identities combine to give a vanishing spacetime covariant divergence to the Einstein tensor, i.e.

$$G_{\mu\nu}{}^{;\nu} = 0$$

$$\Lambda g_{\mu\nu}{}^{;\nu} = 0 \tag{4.1}$$

Therefore,

$$T_{\mu\nu}{}^{;\nu} = 0 \tag{4.2}$$

This is no longer true when we have a local quintessent field. Assuming still zero torsion and metricity, gives at the very least

$$T_{\mu\nu}{}^{;\nu}(x) = -\frac{c^4}{8\pi G}\frac{\partial \Lambda(x)}{\partial x^\nu} g_{\mu\nu}(x) \neq 0 \tag{4.3}$$

---

[301] Lower indices of connection $\Gamma^{\sigma}_{\mu\nu}$ for parallel transport of vectors along paths in Curve World ar symmetric.

[302] Spacetime covariant derivative of metric tensor vanishes.

This is an equation of the "vacuum propeller"[303] type. Torsion and nonmetricity from the universe's "unseen dimensions" would, if present, give further terms for "propellantless propulsion".

Is the quantum vacuum's virtual superfluid charged?

Tony Smith wonders if the infrared virtual tachyon macroscopic quantum vacuum superfluid (Bose-Einstein condensate) whose local phase modulation gives the classical Einstein field, and whose local amplitude modulation gives the quintessent field for both antigravity exotic matter and gravitating dark matter, is charged? That is, is there a new U(1) internal symmetry (hyperspace) group with spontaneous broken local gauge invariance in the vacuum? Is this simply the "axion"? What about the other charges like electro-weak-strong charges (12 altogether)?

I have been rather cavalier on this important detail focusing for the nonce on the delicious results i.e. a unified simple explanation for

1. Why classical curved spacetime at all?

2. Why Einstein's 1915 field equation with zero cosmological constant works so well?

3. Why the universe is accelerating and exotic matter for star gates and warp drive are the same thing on different scales.

4. What the "dark energy" missing mass of the universe really is.

The key new feature is the quantum vacuum coherence factor

$$\Psi(x) = \sqrt{\frac{1}{L_p^3}\left(1 - L_p^2 \Lambda(x)\right)}\, e^{i\Theta(x)}$$

$$(5.1)$$

If $\Psi(x)$ were an absolutely charge neutral giant "wave function" for the macroscopic Bose-Einstein condensate we would not be that well off because we would have no hope of locally controlling it to fly away in our

---

[303] Roger Coolidge (private communication).

saucers through star gates to other worlds of promise, hope and glory. We would be stuck on this small planet only to all die from ecological catastrophe, perhaps quite soon, as many people have described with strange gusto. This is because, we recall (3.1), Einstein's local field for Curve World is

$$g_{\mu\nu}(x) = \frac{1}{2}L_p^2\{\tilde{D}_\mu, \tilde{D}_\nu\}\Theta(x) = \frac{1}{2}L_p^2\left[\tilde{D}_\mu\tilde{D}_\nu + \tilde{D}_\nu\tilde{D}_\mu\right]\Theta(x)$$

(5.2)

where $L_p^2 = hG/c^3 = 10^{-66}$ cm^2 = area of one quantum gravity Bekenstein BIT of Shannon entropy-Brillouin information (depending how you look at it).

"Area" is fundamental to the "world hologram" of my old Cornell chum Lenny Susskind and to "loop quantum gravity" of Ashtekar et-al as a non-perturbative Diff(4) invariant strategy unlike superstring theory with hyperspace which seems to be stuck in perturbation theory, except perhaps for various "dualities"?

If $\tilde{D}_\mu$ is simply a partial derivative in global Flat World, The Capital City of Special Relativity, we do not have much hope of becoming Super Heroes like Buckaroo Banzai, http://jerseyguy.com/bonzai.html rock star (will a Gilbert and Sullivan tenor do?) genius physicist leaping across the 11 dimensions of Super Cosmos. It would be time for me to take off my Captain Video ring that I got at the US Army Quarter Masters Lab in 1950 in lower Manhattan near the John Wanamaker Building.

To make the problem simpler without losing the essence: start in quantum Flat World. Assume Feynman's quantum electrodynamics is a complete description of Flat World. We have only spin 1-boson photons and spin 1/2 fermionic electrons and positrons. An electron of negative energy moving backward in time is a positron of positive energy moving forward in time.

The second quantized electron-positron local field operator $\hat{\psi}_\sigma(x)$ breaks into a positive frequency and a negative frequency part.

$$\hat{\psi}_\sigma = \hat{\psi}_\sigma^+ + \hat{\psi}_\sigma^-$$
$$\sigma = \uparrow, \downarrow \qquad (5.3)$$

The negative frequency part $\hat{\psi}_\sigma^-(x)$, let us say, destroys an electron e^- at spacetime point event x or creates a positron e^+, both of positive energy moving forward in time. Therefore, the positive frequency part $\hat{\psi}_\sigma^+(x)$ creates an electron or destroys a positron. These fermions also have two spin states and the electron-positron complex is described by a 4-component Dirac spinor, which at low energies limits to a 2-component Pauli spinor describing only the electron. I mean here, of course, real quanta on the mass shell!

Is the virtual tachyon superfluid inside the physical quantum vacuum a composite of virtual photons and virtual electron-positron pairs?

First, what about virtual photons forming Bose-Einstein condensates, will that do? NO! Why? Because those are what real detectable classical near induction electric and magnetic fields from motors and transformers etc already are. They are "zero" in the complete vacuum, unless we mean by "vacuum" the electron vacuum? The electron quantum vacuum is an electrically neutral plasma of locally random virtual electron-positron fluctuations. If the spin 0 boson tachyon condensate is composite and not a fundamental field on its own, then it must be a Bose-Einstein condensate of nonlocally connected or "entangled" Einstein-Podolsky-Rosen-Bohm (EPRB) virtual electron-positron pairs like in a BCS superconductor. Yes, that's what it is all right. That is we have a huge macroscopic density of $|\psi(x)|^2 = L_p^{-3}\left[1 - L_p^2 \Lambda(x)\right]$ virtual electron positron pairs all occupying the same small phase space volume, the same nonlocally entangled pair state.

What is the x in the local order parameter $\psi(x)$? It is the center of mass coordinate of the virtual electron-positron pair. How many of these virtual pairs?

Starting with Einstein's 1915 local field equation for the shape of Curve World:

$$G_{\mu\nu}(x) + \Lambda(x) g_{\mu\nu}(x) = -8\pi \frac{G}{c^4} T_{\mu\nu}(x)$$

$$\Lambda(x) \to 0$$

(5.4)

Implies $\psi(x) \neq 0$, i.e. Curve World is the smooth locally nonrandom macroscopic coherent quantum "classical looking" final state from the initial discontinuous locally random microscopic incoherent zero point virtual electron-positron pair vacuum fluctuations. What we have here is a second order phase transition of Goldstone's spontaneous broken symmetry from Flat World to Curve World in which the latter is a smooth nonrandom modulation of the former random carrier channel. Flat World is intrinsically "exotic" with a huge antigravity that will not support ordinary spacetime geometry and matter as we know it.

We then have for $\Lambda \to 0$ to first approximation

$$|\psi|^2 \approx \frac{1}{L_p^3} \approx 10^{99} \, cm^{-3}$$

(5.5)

virtual electron-positron pairs all occupy the *same* entangled pair state in phase space. Use the non positive definite Wigner phase space "wavelet" density $W(x,p)$ whose marginal integrals give $|\psi(x)|^2$ and its Fourier transform $|\tilde{\psi}(p)|^2$ both positive definite.

Each quantum has 4 space-time degrees of freedom and 4 momentum degrees of freedom (tangent fiber bundle). The dimension of phase spacetime is 32 including the 2 spin polarizations for each quantum. However, we will focus on the local center of mass degrees of freedom in 3-space, not the relative degrees of freedom of separation of one fermion from the other in the nonlocally entangled state. The volume of this phase space cell for the center of mass only is $\Box 2\hbar^3$, where $\hbar \sim 10^{-27}$ erg-seconds. Therefore, we have $\sim 10^{99}$ virtual electron-positron pairs per cubic centimeter of spacelike surface all squeezed into this single

"Prospero's Cell" in order to maintain the Curve World in which we move like fish swim in water.

When the density of virtual pairs gets too large we have gravitating electromagnetically invisible "dark energy", i.e. the missing mass of the universe. When the density of virtual pairs gets too small we have antigravitating electromagnetically invisible "exotic matter" (Kip Thorne) needed to accelerate the universe, hold open star gates like Hercules supporting the world and zip around in flying saucers like Buckaroo Banzai and his gang!

http://stardrive.org/cartoon/coffee.html

Consider a nonlocal pair entanglement operator[304]

$$\langle 0| \frac{1}{\sqrt{2}}\left[\hat{\psi}_\uparrow^+(x)\hat{\psi}_\downarrow^-(x') + \hat{\psi}_\uparrow^-(x)\hat{\psi}_\downarrow^+(x')\right]|0\rangle = \frac{1}{\sqrt{2}}\left[e_\uparrow^-(x)e_\downarrow^+(x') + e_\uparrow^+(x)e_\downarrow^-(x')\right]$$
$$(5.6)$$

There is nothing unique about this particular choice of a nonlocal entanglement e-bit pattern because we do not need to consider the electron and the positron identical in the sense of the Pauli exclusion principle. Also, in this case they are in different spin states and can occupy the same event x even if they were identical. This particular choice gives an electrically neutral spin zero complex even on the mass shell for a real electron-positron pair. We will stay off the mass shell. Imagine, for example, 10^{99} of such pairs all Bose-Einstein condensed into this same pair state in each cubic centimeter of classically empty space. To be more precise we need to focus on the center of mass coordinate X of a single pair. All the center of mass coordinates of all the pairs inside the Bose-Einstein condensate are phase-locked together by the phase field $\theta(X)$ of

---

[304] This is one member of a "Bell basis" used in quantum computing, cryptography and teleportation in the Menage a Trois "voyeur" games that Alice, Bob and Eve play with each other in Liaisons Dangereuses. ;-)

the local, but long-range phase coherent, quantum vacuum order parameter $\Psi(X)$

$$X \equiv \frac{x+x'}{2}$$

$$\chi \equiv x - x' \qquad (5.7)$$

The inverse transformation is

$$x = X + \frac{\chi}{2}$$

$$x' = X - \frac{\chi}{2} \qquad (5.8)$$

$$\frac{\partial}{\partial x} = \frac{\partial X}{\partial x}\frac{\partial}{\partial X} + \frac{\partial \chi}{\partial x}\frac{\partial}{\partial \chi} = \frac{1}{2}\frac{\partial}{\partial X} + \frac{\partial}{\partial \chi}$$

$$\frac{\partial}{\partial x'} = \frac{\partial X}{\partial x'}\frac{\partial}{\partial X} + \frac{\partial \chi}{\partial x'}\frac{\partial}{\partial \chi} = \frac{1}{2}\frac{\partial}{\partial X} - \frac{\partial}{\partial \chi} \qquad (5.9)$$

$$\frac{\partial}{\partial x} + \frac{\partial}{\partial x'} = \frac{\partial}{\partial X} \qquad (5.10)$$

Note the cancellation of the relative nonlocal separation in (5.10).

The virtual off mass shell electrons and positrons are certainly charged and they require gauge covariant derivatives in Flat World. That is, we need the local quantum operators

$$-i\vec{\nabla} - \frac{e}{\hbar c}\vec{A}(x) \;\&\; i\frac{\partial}{\partial t} - \frac{e}{\hbar c}A_0(x) \qquad (5.11)$$

Classically the total momentum of an electron in an EM field is $p_\mu - (e/c)A_\mu$ where $-(e/c)A_\mu$ is the EM field momentum "stuck" to the electron. However, in quantum mechanics $\vec{p} \rightarrow (\hbar/i)\vec{\nabla} \ \& \ E \rightarrow i\hbar \partial/\partial t$ as noted above.

The *nonlocal* nonrelativistic[305] quantum Schrodinger equation in configuration space for two charged particles in an external classical electromagnetic field is

$$i\hbar \left\{ \left(\frac{\partial}{\partial t} - \frac{e}{\hbar} A_0(x) \right) + \left(\frac{\partial}{\partial t} - \frac{e}{\hbar} A_0(x') \right) \right\} \psi(x,x')$$

$$= \left\{ -\frac{\hbar^2}{2m} \left[\left(\vec{\nabla}_x - \frac{e}{\hbar c} \vec{A}(x) \right)^2 + \left(\vec{\nabla}_{x'} - \frac{e}{\hbar c} \vec{A}(x') \right)^2 \right] \right\} \psi(x,x')$$

(5.12)

Where $\psi(x,x')$ is the Einstein-Podolsky-Rosen-Bohm nonlocally entangled pair state. What we want to do, quite obviously, is to separate out the center of mass degree of freedom of this single pair. We then have 10^{99} virtual electron-positron pairs per cubic centimeter Bose-Einstein condense, i.e. occupy this same center of mass wave packet spread through ordinary Curve World. The center of mass quantum phase is then coherently locked in step. This is an obvious violation of local gauge invariance, which means that the quantum phases at different events x in Curve World are not tightly locked together. That we have smooth classical Curve World at all would not be able to happen were it not for this phase locking spontaneous broken U(1) symmetry of the virtual off mass-shell electron-positron pairs in a spin-zero electrically neutral macroscopically occupied pair state at the above enormous density if the Planck scale L_p is really as small as 10-33 cm. My theory here works even if L_p is larger as it is in some versions of O Brane New World of M-Theory.[306] Bill Page has cited a paper that large Planck scales from large extra compactification scales in the bosonic Kaluza-Klein sector of the

---

[305] Galilean relativity with Newtonian absolute simultaneity time t.
[306] August 2000 Scientific American, "The Universe's Unseen Dimensions" and Stephen Hawking's "The Universe in a Nutshell".

hyperspace of Super Cosmos would conflict with data on proton decay. I am not sure about that.

The rule in Curve World's *configuration space* is therefore

$$D_\mu \to D_\mu(x) + D_\mu(x')$$

(5.13)

The problem here in Curve World is that curvature[307] and torsion[308] introduce non-integrable path-dependences needing the connection fields. The world line one takes in parallel transport matters. However, here in Quantum World we are in a configuration space and at this point the anholonomies of local Curve World do not yet impede the progress of your Journey with me along the straightest hyperspace geodesic through Dante's Inferno on the Path of Enlightenment to Paradiso and the City of the Mind of God.[309] ☺

$$\{D_\mu(x) + D_\mu(x')\}\psi(x,x') = \left\{-i_{(\mu)}\frac{\partial}{\partial x^\mu} - \frac{e}{\hbar c}A_\mu(x) - i_{(\mu)}\frac{\partial}{\partial x'^\mu} - \frac{e}{\hbar c}A_\mu(x')\right\}\psi(x,x')$$

(5.14)

$$\{D_\mu(x) + D_\mu(x')\}\psi(x,x') = \left\{-i_{(\mu)}\frac{\partial}{\partial X^\mu} - \frac{e}{\hbar c}\left[A_\mu\left(X + \frac{\chi}{2}\right) + A_\mu\left(X - \frac{\chi}{2}\right)\right]\right\}\psi'(X,\chi)$$

(5.15)

---

[307] Disclination string defects in the World Crystal Lattice of Hagen Kleinert.

[308] Dislocation string defects breaking closed paths into open paths with a gap in the local Curve World <-> Flat World "tetrad map" that is Einstein's local principle of equivalence, i.e. explain gravity by locally eliminating gravity in the free float weightless local inertial frames (LIF) of Flat World tangent fiber space. Frames of reference stuck in Curve World are non-inertial, i.e. non-geodesic with "weight" from non-gravity electrical reaction forces. For example, standing on the surface of the Earth is a noninertial frame in Curve World. Weightless free float on the Space Shuttle in orbit with engines off is an LIF frame with zero gravity.

[309] See end of Hawking's "A Brief History of Time".

I now make the usual Ansatz of separability. This needs more rigor of course, but I leave that for the mopping up by future grad students doing dissertations at Star Fleet Academy in San Francisco's Presidio.

$$\psi'(X,\chi) = \Psi(X)\psi(\chi) \tag{5.16}$$

Therefore

$$\{D_\mu(x) + D_\mu(x')\}\psi(x,x') \rightarrow \left\{D_\mu\left(X+\frac{\chi}{2}\right) + D_\mu\left(X-\frac{\chi}{2}\right)\right\}\Psi(X)\psi(\chi)$$

$$= \left\{-i_{(\mu)}\frac{\partial}{\partial X^\mu} - \frac{e}{\hbar c}\left[A_\mu\left(X+\frac{\chi}{2}\right) + A_\mu\left(X-\frac{\chi}{2}\right)\right]\right\}\Psi(X)\psi(\chi) \tag{5.17}$$

Next, make a Taylor series expansion on the electromagnetic 4-potentials A_μ. Symbolically and *very* non-rigorously this is

$$A_\mu\left(X\pm\frac{\chi}{2}\right) \approx \sum_{n=0}^{\infty}\frac{(\pm 1)^n}{n!}\left(\frac{\chi}{2}\right)^2\left(\frac{d}{dX}\right)^n A_\mu(X) \tag{5.18}$$

This is really a multiple Taylor expansion of course since each variable is a set of 4 variables. What we have here is essentially an electromagnetic multipole expansion. We are concerned here with the post-quantum back action BIT FROM IT covariant Landau-Ginzburg equation for the deformation of the Bose-Einstein condensate by Curve World given that Curve World itself is an emergent coherent collective order[310] from the phase modulation of the Bose-Einstein condensate itself. What we have here is illustrated in Maurice Escher's "Drawing Hands"

---

[310] P.W. Anderson's "More Is Different".

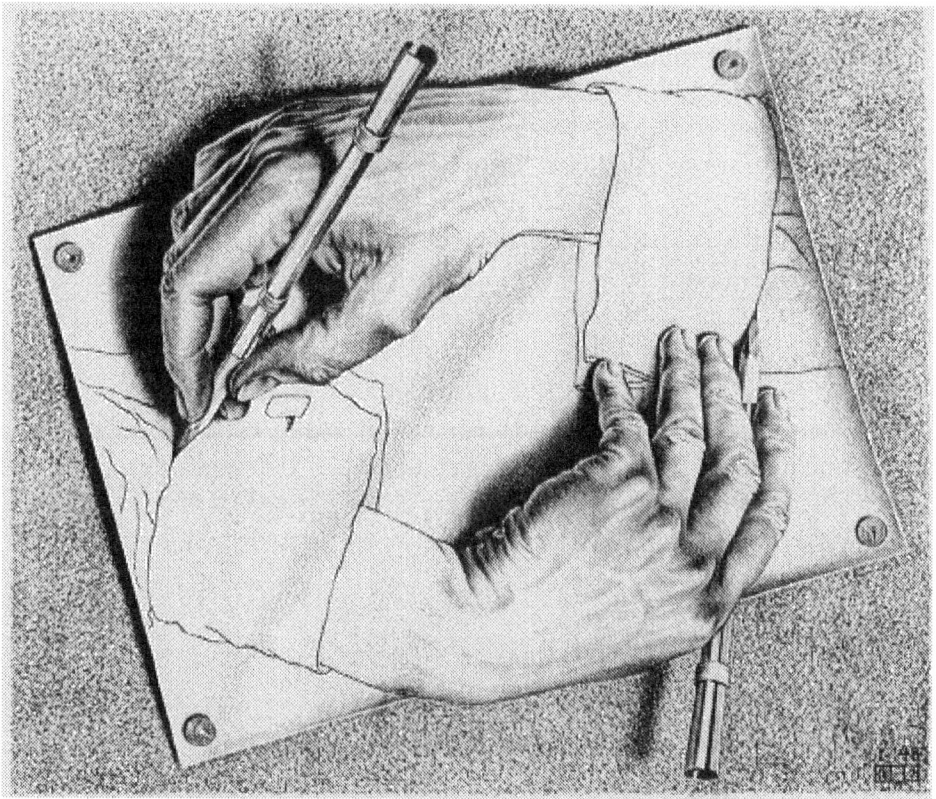

Note that the odd multipoles cancel out of the center of mass motion equation. For now, restrict the pair separation χ to be small compared to local radii of spacetime curvature. If the Landau-Ginzburg equation were linear the $\psi(\chi)$ function would cancel out. But it is nonlinear, hence it does not cancel out. Thus we must also take the difference

$$\{D_\mu(x) - D_\mu(x')\}\psi(x,x') \to \left\{D_\mu\left(X + \frac{\chi}{2}\right) - D_\mu\left(X - \frac{\chi}{2}\right)\right\}\Psi(X)\psi(\chi)$$

$$\left\{-2i_{(\mu)}\frac{\partial}{\partial\chi^\mu} - \frac{e}{\hbar c}\left[A_\mu\left(X + \frac{\chi}{2}\right) - A_\mu\left(X - \frac{\chi}{2}\right)\right]\right\}\Psi(X)\psi(\chi)$$

$$(5.19)$$

285

Now it is the even multipoles that cancel out. The complete nonlinear BIT FROM IT post-quantum backaction partial differential equation will be a set of two coupled equations in which the center of mass motion and the internal motion of the pair state are mutually interdependent.

We really want an equation only for the center of mass of the pair with the effective potential of (3.7) above. When in doubt, integrate out!

What we want to do is to take 3-dimensional spacelike integrals over the relative separation χ. This is like integrating over unobserved particles in the reduced density matrix formalism of many-particle physics. In particular look at the nonlinear and nonlocal potential terms of the Landau-Ginzberg equation in the configuration space of the virtual pair. Imagine that the basic equation in Curve World virtual particle-antiparticle configuration space is

$$D^{\mu}\left(x,x'\right)D_{\mu}\left(x,x'\right)\psi\left(x,x'\right)=0 \tag{5.20}$$

That is,

$$D^{\mu}\left(X,\chi\right)D_{\mu}\left(X,\chi\right)\Psi\left(X\right)\psi\left(\chi\right)=0 \tag{5.21}$$

Take the symmetric form,

$$D\left(X,\chi\right)_{s} \equiv D\left(X+\frac{\chi}{2}\right)+D\left(X-\frac{\chi}{2}\right)$$

$$=-i_{(\mu)}\frac{\partial}{\partial X^{\mu}}-\frac{2e}{\hbar c}\left[A_{\mu}\left(X+\frac{\chi}{2}\right)+A_{\mu}\left(X-\frac{\chi}{2}\right)\right] \tag{5.22}$$

Next, form the symmetric wave propagation operator in pair configuration space[311]

---

[311] We can leave out the Curve World Γ terms since they do not contribute to the effective potential for the order parameter $\Psi\left(X\right)$.

$$D(X,\chi)^\mu_{\ s}\, D(X,\chi)_{\mu s} \approx$$

$$g^{\mu\sigma}(X)\left\{-i_{(\sigma)}\frac{\partial}{\partial X^\sigma}-\frac{2e}{\hbar c}\left[A_\sigma\left(X+\frac{\chi}{2}\right)+A_\sigma\left(X-\frac{\chi}{2}\right)\right]\right\}\left\{-i_{(\mu)}\frac{\partial}{\partial X^\mu}-\frac{2e}{\hbar c}\left[A_\mu\left(X+\frac{\chi}{2}\right)+A_\mu\left(X-\frac{\chi}{2}\right)\right]\right\}$$

$$(5.23)$$

Now integrate all χ-dependent terms in the wave operator with respect to a spacelike slice of spacetime. The basic conditional probability integral operator for these terms is

$$\iiint d^3\chi\left|\psi(\vec{\chi},\chi_0)\right|^2 \dots \qquad (5.24)$$

Obviously the pure center of mass differential operator terms $-i_{(\sigma)}\dfrac{\partial}{\partial X^\sigma}$ are not affected. It seems obvious to me that we will pull out the effective potential of (3.7) in this process in which the parameter M^2, which can be positive or negative, comes from the χ integrations (5.24) over the gauge force multipole expansion of (5.18), and β comes from the χ integrals in the screened Hartree-Fock Coulomb static field approximation. This screened field potential will be repulsive. Why? There are three contributions of essentially equal strength in the virtual electron-positron charge neutral plasma that is the quantum vacuum of the Dirac field. They are electron-electron, positron-positron, and electron-positron. That is two repulsive vs. one attractive interaction, hence repulsion dominates which is exactly what we want. The center of mass densities; $\left(\Psi^*(X)\Psi(X)\right)^2$ are outside the χ integrals in the screened Hartree-Fock Coulomb static field computation. Therefore,

$$\left(\frac{Mc}{\hbar}\right)^2=\left(\frac{2e}{\hbar c}\right)^2 g^{\mu\sigma}(X)\iiint d^3\chi\left[A_\sigma\left(X+\frac{\chi}{2}\right)+A_\sigma\left(X-\frac{\chi}{2}\right)\right]\left[A_\mu\left(X+\frac{\chi}{2}\right)+A_\mu\left(X-\frac{\chi}{2}\right)\right]$$

$$(5.25)$$

$$-\infty < M^2 < +\infty \qquad (5.26)$$

$$\frac{\beta}{\hbar^2}\left(\Psi^*(X)\Psi(X)\right)^2$$

$$\approx e^2 \iiint d^3\chi \left\{ \frac{\left[\rho_1\left(X+\frac{\chi}{2},X+\frac{\chi}{2}\right)\rho_1\left(X-\frac{\chi}{2},X-\frac{\chi}{2}\right)+2\rho_1\left(X+\frac{\chi}{2},X-\frac{\chi}{2}\right)^2\right]}{|\vec{\chi}|} \right\}$$

$$(5.27)$$

Using the reduced density matrix formalism. The off-diagonal term is the quantum mechanical Heisenberg, in this case repulsive, Coulomb exchange interaction between virtual electrons with virtual electrons on different pairs, and also between virtual positrons with virtual positrons on different pairs. The Coulomb attraction between virtual electrons with positrons in the same pair and on different pairs cancel out. There, is of course, no exchange interaction in that case since they are not to be treated as identical particles in the sense of the Pauli exclusion principle.

$$\rho_1\left(X+\frac{\chi}{2},X+\frac{\chi}{2}\right) \equiv \iiint d^3x'\left|\psi(x,x')\right|^2$$

$$\rho_1\left(X-\frac{\chi}{2},X-\frac{\chi}{2}\right) \equiv \iiint d^3x\left|\psi(x,x')\right|^2$$

$$(5.28)$$

$$\rho_1\left(X+\frac{\chi}{2},X-\frac{\chi}{2}\right) \equiv \left|\psi(x,x')\right|^2$$

$$(5.29)$$

The effective potential for the quantum vacuum order parameter $\Psi(X)$ whose coherent phase modulation is Einstein's classical curved spacetime geometrodynamic field $g_{\mu\nu}(x)$ is

$$V\left(\Psi^*(X),\Psi(X)\right) = Mc^2\Psi^*(X)\Psi(X)+\frac{\beta}{M}\left(\Psi^*(X)\Psi(X)\right)^2$$

$$(5.30)$$

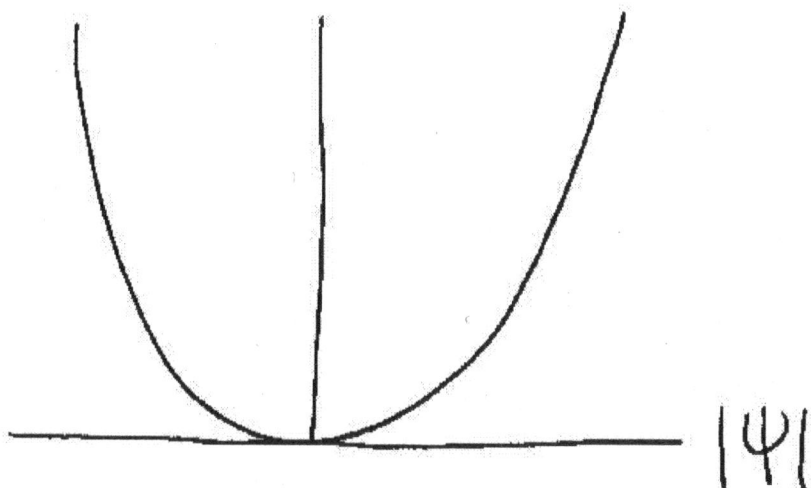

FALSE VACUUM
OF FLAT WORLD
$$M^2 > 0 \, , \, \beta > 0$$

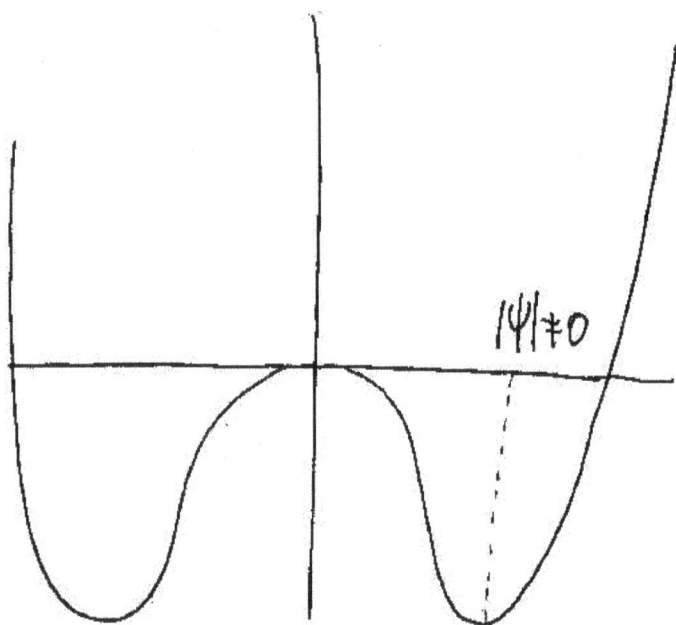

$|\psi| \neq 0$

TRUE VACUUM
OF CURVE WORLD
$M^2 < 0, \beta > 0$

EFFECTIVE POTENTIAL

MASSIVE
HIGGS
OSCILLATIONS
OF AMPLITUDE
OF ψ

GAPLESS
GOLDSTONE
OSCILLATIONS
OF PHASE
OF ψ

ψ = VIRTUAL TACHYON CONDENSATE
BOSE - EINSTEIN ▪▪▪▪▪▪

EINSTEIN'S GRAVITY IS FROM
MODULATING PHASE OF ψ

THE BASE IS THE COMPLEX PLANE

$$\frac{M}{\hbar^2}\frac{\delta V(\Psi^*,\Psi)}{\delta \Psi^*}=\left(\frac{Mc}{\hbar}\right)^2\Psi+\frac{\beta}{\hbar^2}|\Psi|^2\Psi\to 0$$

$$|\Psi|^2=-\frac{(Mc)^2}{\beta}$$

$$M^2<0 \tag{5.31}$$

Is the more stable vacuum at the minimum of the effective potential. The zero root is now an unstable vacuum of higher energy density. Use equations (5.25) and (5.27), symbolically

$$\frac{(Mc)^2}{\beta}=\frac{\left(\frac{2e}{c^2}\right)^2 g^{\mu\sigma}(X)\iiint d^3\chi\left[A_\sigma\left(X+\frac{\chi}{2}\right)+A_\sigma\left(X-\frac{\chi}{2}\right)\right]\left[A_\mu\left(X+\frac{\chi}{2}\right)+A_\mu\left(X-\frac{\chi}{2}\right)\right]}{|\Psi|^4 e^2 \iiint d^3\chi\left\{\dfrac{\left[\rho_1\left(X+\frac{\chi}{2},X+\frac{\chi}{2}\right)\rho_1\left(X-\frac{\chi}{2},X-\frac{\chi}{2}\right)+2\rho_1\left(X+\frac{\chi}{2},X-\frac{\chi}{2}\right)^2\right]}{|\vec{\chi}|}\right\}}$$

$$=-|\Psi|^2 \tag{5.32}$$

$$|\Psi|^2=\left[\frac{-\left(\frac{2e}{c}\right)^2 g^{\mu\sigma}(X)\iiint d^3\chi\left[A_\sigma\left(X+\frac{\chi}{2}\right)+A_\sigma\left(X-\frac{\chi}{2}\right)\right]\left[A_\mu\left(X+\frac{\chi}{2}\right)+A_\mu\left(X-\frac{\chi}{2}\right)\right]}{e^2\iiint d^3\chi\left\{\dfrac{\left[\rho_1\left(X+\frac{\chi}{2},X+\frac{\chi}{2}\right)\rho_1\left(X-\frac{\chi}{2},X-\frac{\chi}{2}\right)+2\rho_1\left(X+\frac{\chi}{2},X-\frac{\chi}{2}\right)^2\right]}{|\vec{\chi}|}\right\}}\right]^{-\frac{1}{3}}$$

$$=\frac{1}{L_p^3}\left[1-L_p^2\Lambda(X)\right] \tag{5.33}$$

The local quintessent field $\Lambda(X)$ is, therefore, determined implicitly in terms of quantum electrodynamics since it is essentially a coherence of virtual electron-positron pairs. The algebra is complicated and non-intuitive so I have to check to make sure I have not left out an explicit dependence on the spacetime connection Γ. Remember this is the first time, on this planet, at least, that ideas like this have ever been written down. I am rather amazed at what I am doing myself. This was unexpected even two days ago, but was cosmically triggered by a query from Tony Smith on which U(1) I meant. I mean electromagnetic U(1). Note also I do not need torsion, nor do I need large extra-space dimensions to explain key mysteries about our universe in a simple way. The ideas are simple, only the algebra is a bit complicated but with Math Type who cares? ;-) However, I can add torsion and extra large space dimensions rather easily, but so far no need.

Star Gate Time Travel Metric
Kip Thorne's toy static[312] spherically symmetric metric for a nonsingular traversable wormhole time machine has the generic form[313]

$$ds^2 = -e^{+2\phi_\pm(r)}c^2dt^2 + \frac{1}{\left\{1 - \frac{b_\pm(r)}{r}\right\}}dr^2 + r^2 d\Omega$$

(6.1)

$\phi_\pm(r)$ are the two redshift functions since we will need two separate coordinate patches for the two mouths or portals of the Star Gate. The two mouths may be in the same universe (3-brane) at same or different global

---

[312] "static" = time independent (stationary) and nonrotating, symmetric under time reversal & global Killing vector field that is hypersurface orthogonal and timelike in limit of spacelike infinity. Lie dragging derivative of metric along Killing vector field vanishes by definition (isometry).

[313] Matt Visser and John Peacock use different GR sign conventions that may cause me to make a sign error.

cosmic times[314], or they may connect parallel universes separated from each other in the hyperspace of Super Cosmos[315].

When

$$\phi_+ (\infty) \neq \phi_- (\infty) \tag{6.2}$$

We have a time machine. Moving through the Star Gate one way takes you to the future. Moving the opposite way takes you to the past. Stephen Hawking has a "chronology protection conjecture" that says you will be burned to a crisp by hard blue shifted radiation as soon as this time travel condition emerges. I think Hawking is wrong because his argument is based on flimsy "renormalization" arguments that Feynman called a "scandalous shell game" even though he was one of its main inventors. I suspect my new quintessent field $\Lambda(x)$ will show the error in Hawking's argument. I could be wrong here of course. But damn the torpedoes, let's see what we get if we do not live in the boring universe Hawking leaves us with.

$b_\pm (r)$ is the shape function[316] of the traversable wormhole with no event horizon where time comes to a stop and no black hole spacetime singularity behind the event horizon to stretch and squeeze us to death, even our atoms out of existence in the Devil's Rack!

$$b_\pm \equiv b_\pm (\infty) = \frac{2GM_\pm}{c^2} \tag{6.3}$$

---

[314] Measured by absolute temperature of the black body thermal radiation from the end of the Big Bang.

[315] "Universe in a Nutshell" by Stephen Hawking, "Hyperspace" by Michio Kaku, "The Universe's Unseen Dimensions" August 2000, Scientific American. Tiny proton decay rate may make big hyperspace compactification scale impossible? (Bill Page e-mail).

[316] We are using Schwarzschild coordinates. The radial derivatives of both the redshift and shape functions match at the minimum of the throat of the wormhole. Details in Visser 11.2 of "Lorentzian Wormholes".

Using the quintessent field for a pure exotic quantum vacuum engineered Star Gate

$$M_{\pm} = -\frac{\Lambda(R_{\pm})c^2 R_{\pm}^3}{2G}$$

(6.4)

Where $R_{\pm}$ is the scale of the Star Gate "entrance-exit" or "doorway" Ad Astra[317] and Beyond in the Secret Passage to the New India, the fabled land of "Magonia". I make the approximation that the quintessent field $\Lambda(R_{\pm})$ is uniform over the scale of the wormhole mouth in each parallel universe[318], or in the two entrance and exit events of the same universe as the case may be.

For this simple toy model Star Gate from Cal Tech, the Einstein tensor components are few and have the generic form

---

[317] "To The Stars" Royal Air Force motto in the Battle of Britain in WWII against Hitler and Goring's Luftwaffe. Today we have "The Axis of Evil", al Qaida.

[318] I mean "parallel universe" in the classical material IT sense of Bohm's hidden variable system point moving on the mental qubit pilot wave landscape or BIT in sense of Wheeler's IT FROM BIT. David Deutsch's multiverse, is something else. It refers to the different BIT valleys, or attractor basins for IT the not so Hidden Variable "system point", on the grand mental landscape of Super Cosmos i.e. Hawking's "Mind of God".

$$G_{\pm tt}(r) = \frac{1}{r^2} \frac{\partial b(r)_{\pm}}{\partial r}$$

$$G_{\pm rr}(r) = -\frac{b(r)_{\pm}}{r^3} + 2\left\{1 - \frac{b(r)_{\pm}}{r}\right\}\frac{1}{r}\frac{\partial \phi(r)_{\pm}}{\partial r}$$

$$G_{\pm \theta\theta}(r) = G_{\pm \varphi\varphi}(r) = \left\{1 - \frac{b(r)_{\pm}}{r}\right\}\left[\frac{\partial^2 \phi(r)_{\pm}}{\partial r^2} + \frac{\partial \phi(r)_{\pm}}{\partial r}\left(\frac{\partial \phi(r)_{\pm}}{\partial r} + \frac{1}{r}\right)\right]$$

$$-\frac{1}{2r^2}\left[\frac{\partial b(r)_{\pm}}{\partial r}r - b(r)_{\pm}\right]\left(\frac{\partial \phi(r)_{\pm}}{\partial r} + \frac{1}{r}\right)$$

$$(6.5)$$

My approximate field equation by passing the spacetime stiffness barrier of 1 fermi[319] bend per 4 billion metric tons equivalent applied external electromagnetic energy density[320] is,

$$G_{\mu\nu} = -\Lambda g_{\mu\nu} \tag{6.6}$$

Define

$$\rho(r) \equiv \frac{\Lambda c^2}{8\pi G} g_{tt}(r) = -\frac{\Lambda(r)c^2}{8\pi G}e^{+2\phi_{\pm}(r)}$$

$$-\tau \equiv \frac{\Lambda c^2}{8\pi G} g_{rr} = \frac{\Lambda c^2}{8\pi G}\frac{1}{\left\{1 - \dfrac{b_{\pm}(r)}{r}\right\}}$$

$$p \equiv \frac{\Lambda c^2}{8\pi G} g_{\theta\theta} \equiv \frac{\Lambda c^2}{8\pi G} g_{\varphi\varphi} = \frac{\Lambda c^2}{8\pi G} \tag{6.7}$$

Substitute (6.1) and (6.5) into (6.6) to get the IT FROM BIT equations:

---

[319] One thousandth of a one billionth of one centimeter! (fermi)

[320] This is why it is completely foolish of Eric Davis and Hal Puthoff to use a theory of UFOs limited by such a barrier. There is a way around it as I think I am showing here.

$$G_{\pm tt}(r) = \frac{1}{r^2}\frac{\partial b(r)_{\pm}}{\partial r} = \Lambda(r)e^{+2\phi_{\pm}(r)}$$

$$G_{\pm rr}(r) = -\frac{b(r)_{\pm}}{r^3} + 2\left\{1 - \frac{b(r)_{\pm}}{r}\right\}\frac{1}{r}\frac{\partial \phi(r)_{\pm}}{\partial r} = -\frac{\Lambda(r)}{\left\{1 - \frac{b_{\pm}(r)}{r}\right\}}$$

$$G_{\pm\theta\theta}(r) = G_{\pm\varphi\varphi}(r) = \left\{1 - \frac{b(r)_{\pm}}{r}\right\}\left[\frac{\partial^2 \phi(r)_{\pm}}{\partial r^2} + \frac{\partial \phi(r)_{\pm}}{\partial r}\left(\frac{\partial \phi(r)_{\pm}}{\partial r} + \frac{1}{r}\right)\right]$$

$$-\frac{1}{2r^2}\left[\frac{\partial b(r)_{\pm}}{\partial r}r - b(r)_{\pm}\right]\left(\frac{\partial \phi(r)_{\pm}}{\partial r} + \frac{1}{r}\right) = -\Lambda(r)$$

$$(6.8)$$

This is only half the story of course; we still have the BIT FROM IT post-quantum back-action Landau-Ginzberg equation and Maxwell's field equations in Curve World to deal with. I never promised you a Rose Garden. It will probably take supercomputing to get engineering results unless some young genius comes along with simple analytical models?

What do the UFOs do?

http://www.broadwaymidi.com/down/Camelot-WhatDoTheSimpleFolkDo.mid

Eric Davis in
http://198.63.56.18/pdf/davis_mufon2001.pdf

makes a big deal that Hal Puthoff's PV theory explains the alleged property of real UFO's described as "fiction" in Fastwalker by Jacques Vallee and Tracy Torme (Mel's son) writer producer of TV series Sliders on parallel universes.

http://www.nidsci.org/bios/vallee.html

This is no big deal.

In http://stardrive.org/Jack/Casimir.pdf [321]

One sees that the effective potential per unit mass of the quantum vacuum in the weak field limit obeys the Poisson equation[322]

$\nabla^2 V$ = -4πG(effective mass density + 3 pressure/c^2)

with the quantum vacuum equation of state

effective mass density + pressure/c^2 = 0

And

effective mass density = Λ (x) c^2/8πG

Λ (x) (x) is the local quintessent field from the amplitude of the order parameter of the Bose-Einstein condensate of virtual electron-positron pairs whose quantum phase variation gives Einstein's classical geometrodynamic field $g_{\mu\nu}$ (x) of 1915 GR (e.g. Hagen Kleinert's web page for background).

OK take a flying saucer whose outer edge has a hollow ring of square cross section a^2 of radius R for a total volume of 2πRa^2.

There are control induction EM *nonradiating near fields* inside the ring vacuum chamber. The effective active gravitational mass of the quantum vacuum in the ring is then

$$M_{vac} \equiv -\frac{\Lambda(R)c^2 Ra^2}{2G}$$

(7.1)

Redshift z is

$$z \equiv \frac{\lambda_o - \lambda_e}{\lambda_e} \equiv \frac{\Delta\lambda}{\lambda_e}$$

$$\Delta v = \Delta\left(\frac{c}{\lambda_e}\right) = -\frac{c}{\lambda_e^2}\Delta\lambda$$

$$\frac{\Delta v}{v_e} = -z$$

(7.2)

A positive z is a redshift, a negative z is a blueshift, subscript "e" is the emission event. Similarly, "o" is the observation event. Roughly, for these non-classical vacua

$$z = -\frac{\Delta v}{v_e} = \frac{v_e - v_o}{v_e} \approx \frac{\Lambda(R)Ra^2}{2}\left[\frac{1}{(R+r)} - \frac{1}{R}\right]$$

$$\frac{1}{(R+r)} - \frac{1}{R} < 0$$

(7.3)

In the near field r/R < 1, the limit of small r is obviously zero

$$\frac{1}{(R+r)} - \frac{1}{R} = \frac{1}{R\left(1 + \dfrac{r}{R}\right)} - \frac{1}{R}$$

$$= \frac{1}{R}\sum_{n=0}^{\infty}(-1)^n\left(\frac{r}{R}\right)^n - \frac{1}{R} = \sum_{n=1}^{\infty}(-1)^n\left(\frac{r}{R}\right)^n$$

(7.4)

In the far field R/r < 1

$$\frac{1}{(R+r)} - \frac{1}{R} = \frac{1}{r\left(1+\dfrac{R}{r}\right)} - \frac{1}{R}$$

$$= \frac{1}{r}\sum_{n=0}^{\infty}(-1)^{n}\left(\frac{R}{r}\right)^{n} - \frac{1}{R} = \frac{1}{r}\sum_{n=1}^{\infty}(-1)^{n}\left(\frac{R}{r}\right)^{n} - \frac{1}{R}$$

$$\xrightarrow[r\to\infty]{} -\frac{1}{R}$$

$$(7.5)$$

Therefore,

$\Delta v/v_e$ is a blue shift when $\Lambda > 0$

$\Delta v/v_e$ is a redshift when $\Lambda < 0$

It is assumed that $\Lambda = 0$ at the detector which is ordinary as distinct from exotic vacuum.

This quintessent quantum spectral shift is a much bigger effect than one gets from the $(G/c^4)T_{\mu\nu}$ term!

That's what UFOs if real do!

So they say.

Note that

$$\Lambda L_p{}^2 = 1 - L_p{}^3 |\psi|^2$$

Where $|\psi|^2 = 10^{99}$ Bose-Einstein condensed virtual electron-positron spin zero electrically neutral pairs per cubic centimeter.

So what spectral shift do you want Bhubba?

"We spectres are a jollier crew than you, perhaps, suppose." Ruddygore G & S

http://math.boisestate.edu/gas/ruddigore/html/night_wind_howls.html

The Chipman File

The late Harold Chipman was a Chief of Station of the Central Intelligence Agency who briefly funded my work in the mid 1980's. He claimed to have been involved with remote viewing projects behind the scenes including the SRI Remote Viewing by Puthoff and Targ in the early 70's. Hal Puthoff disputes Chip's claims. On the other hand, "Chip" said that Hal was unaware of his role. The situation is a bit like "Rashomon". The following documents attest to Chipman's efforts to set up an SDI-oriented think tank to develop my physics ideas. My ideas in 1985 had not yet gelled. The turning points were in 1994 and 2001 as shown in "Zero Point Energy, Star Gates, and Warp Drive" in this volume.

CORPORATION STRUCTURE

APRIL 1985

Function	Entity	Ownership
Patent Control	SARFATTI CORPORATION California	49% Sarfatti 51% Haenlein
		100% Ownership
Marketing and Licencing	QRDC California	14.25% Haenlein 14.25% Chipman 14.00% Sarfatti 4.00% Levit 2.00% Rudl 1.00% Altenberg 15.00% Phase I Investors 35.50% Phase II Investors
		100.00% Ownership
Think Tank	SARFATTI INSTITUTE California	100.00% Sarfatti Private Corporation
		100.00% Ownership

SARFATTI INSTITUTE

This Institute, proposed to be located at the University
of San Francisco, will be funded by the Sarfatti
Corporation.

GOALS OF THE INSTITUTE

The goals will be:

1. to conduct basic research in the quantum physics
of nonlocal phenomena discovered by Albert Einstein in
1935 (Einstein-Podolsky-Rosen-Bohm effect), and

2. to develop applications of the basic mathematics
and physics research into patents for untappable,
unjammable command-control-communications, sixth
generation quantum spin correlation supercomputers,
quantum zero point energy release for interstellar
rocket engines, supertechnology for the Strategic
Defense Initiative (SDI) to reduce the possibility of
nuclear war, pi-orbital electron spin control of
recombinant genetic engineering, manipulation of
Hoyle-Godel "loops in time", quatitative investigation
of the mind-matter interaction, and other fundamental
applications as they emerge from the basic research
program.

STYLE OF THE INSTITUTE

The style will be:

1. top minds, both established and up-and-coming, should
be paid to come and conduct research in an environment
where they are unafraid to speculate,

2. seminars, both formal and informal, shall take place.
Entertaining educational media shall be developed for
the public,

3. scientists shall be encouraged to debate their
"rivals," and

4. scientists, poets, artists, political theorists,
philosophers, military and intelligence officers,
politicians, captains of industry, theologians, etc,
shall be gathered together for workshops on the cultural
impact of breakthroughs in the "New Physics."

I. THE BUSINESS

The business falls into the category of pure research and development in the supertechnology field. By definition, it is a high risk - high gain endeavor.

Succinctly, Quantum Research Development Corporation (QRDC) has the exclusive rights for marketing and licensing for manufacture all devices resulting from a body of work developed over two decades by an American theoretical physicist. The physicist postulates that new switching devices can be made, rendering obsolete myriad devices in current use affecting computer and communications technology. Further, and more importantly, such devices offer a quantum leap into the super-technology required for the next century.

QRDC is under contract to test these devices in a laboratory with impeccable credentials to certify the results. The second phase involves marketing and licensing. Downstream, QRDC would assist in establishing an independent non-profit institute where the pure physics of the new effect would be researched. QRDC will market the resulting devices invented.

The device to be used in the test, a pair-correlated photon generator, is a basic component of the switches to be manufactured commercially. Ultimately, these "quantum spin" switches would replace the conventional solid state computer chip and other communications electronic elements.

The Developer

The developer/discoverer of the hypothesis is 45 year old, New York born, theoretical physicist, Dr. Jack Sarfatti (PhD), graduate of Cornell, University of California San Diego and Riverside, author of numerous papers and articles on a wide range of articles dealing with theoretical physics. His work and views are known internationally. (See TAB A).

The history of the development of this hypothesis is found in TAB B and Dr. Sarfatti's Statement to Investors is found in TAB C.

Dr. Sarfatti is a founding Board Member and holds 14% interest in QRDC.

FOREWARD

In April 1985, an American theoretical physicist
completed final work on a theory which, if proved in
laboratory tests, alters the fundamental scientific
interpretation of quantum mechanics and our general
understanding of the universe. His calculations represent
the next logical step in the theoretical physics work of
Albert Einstein, Boris Podolsky, Nathan Rosen and David
Bohm.

This new hypothesis provides a quantum view of the
structure of the universe. It offers new clarity to the
quantum theory, advancing it to the next level of
understanding. */

The hypothesis, when tested, has the potential to usher
in that era of supertechnology required by the United States
to maintain the technological superiority of the West into
the next century. The impact of the inventions resulting from
this new approach on the economic, political, military,
intellectual, religious, social and scientific levels of
society, would be profound.

The physicist has also conceptualized the first two
basic devices, representing the practical, economic
application of the hypothesis. Patent work is underway for
both devices.

Quantum Research Development Corporation was organized
in March 1985 to harness the economic and strategic potential
offered.

---

*/ Succinctly, the Sarfatti hypothesis holds that:

> There is a faster-than-light (superluminal)
> action-at-a-distance effect that can be
> harnassed for practical use in communications.

CHIPMAN GROUP LIMITED
SUITE 600
170 MAIDEN LANE
SAN FRANCISCO, CALIFORNIA 94108

LONDON
HONG KONG

TELEX: 176065
GALEBOB-SF

HAROLD E. CHIPMAN
550 Battery Street
San Francisco, California 94111
(415) 362-4148

Professional
Experience

CHIPMAN GROUP LIMITED   1978 - Present
President and Chairman of the Board   San Francisco

Private investment banking. Syndicating domestic and foreign
funds for construction, manufacturing and promotional ventures.
Offices and personnel located in Europe and Asia. Affiliated
with international merchant banking network.

International political and economic action investment ventures.

MARKET ANALYSIS INCORPORATED   1978 - 1982
President   San Francisco

Production of political and economic risk analyses for American
multi-national corporations investing/divesting abroad; marketing
intelligence for American corporations expanding into international markets; information for foreign firms expanding into American
markets.

Secured private financing for American corporations, co-ordinating
off-shore construction and investment projects; provided venture
capital for American inventors.

Conducted seminars for senior officers of Fortune 100 corporations
on current and projected political and economic matters.

Provided analyses and served as ghost writer for certain domestic
and foreign periodicals, newspapers, magazines and journals; wrote
analyses for select domestic political figures.

Directed work of headquarters staff employees and chiefs of corp-
orate divisions:

- Air/Maritime   - Finance
- Commodity Trading   - Precious Metals
- Construction   - Energy Resources

CHIPMAN GROUP LIMITED

PAUL BERRY ASSOCIATES	1978 - Present
Senior Partner	San Francisco

Merchant banking operations.

DOMAIN INTERNATIONAL	1980 - Present
Partner	London and Hamilton, Bermuda

Construction system.

UNITED STATES GOVERNMENT	1950 - 1977

(1) National Security Agency (1950-1954)   Arlington, Virginia

Research analyst specialist, current intelligence writer/editor, linguist and cryptographer.

(2) Foreign Service (1955-1977)   Europe
                                                                   Latin America
                                                                   Africa
                                                                   Asia

Political Officer and Plans Officer with specific interest in the Soviet Union, Eastern Bloc, Cuba and other Marxist states.

Prepared and edited reporting and analyses for national policy makers. Directed widescale activities of individuals and groups engaged in promoting United States foreign policy objectives abroad. Engaged in economic and political action activities to further these policies. Conducted negotiations with senior political and economic leaders abroad.

U.S. Government "Key Official Emergency Relocation Cadre". Last foreign posting: U.S. Consul, Hamburg (Political Officer) 1974.

Maximum number of foreign and American employees: 1,500.

Military Service

UNITED STATES MARINE CORPS	1945 - 1947

Active and reserve duty. Marine Aviation. Retired Reserve 1968.

Education

GEORGETOWN UNIVERSITY	1948 - 1951
School of Foreign Service BSFS, Economics	Washington, DC

GEORGETOWN UNIVERSITY	1951 - 1952
Graduate School Political Science	Washington, DC

CHIPMAN GROUP LIMITED

GEORGE WASHINGTON UNIVERSITY   1952 - 1953
Journalism   Washington, DC

Professional Organizations
Past and Present

    Rotary International   San Francisco
    Propellor Club   New York

Civic and Social Organizations
Past and Present

    Explorers Club   New York
    Ueberseeklub   Hamburg
    German American Club   Hamburg
    Kieler Yacht Klub   Kiel
    American Embassy Club   Bonn
    Boy Scouts of America, Marin County (Staff)   Marin County

Personal

    Marital Status - Divorced
    Excellent Health
    Available for international travel except for Soviet and Bloc States
    Awards -   Commendation Department of Defense (1965)
                     Certificate of Merit, Department of Defense (1963)
                     Vietnamese Police Medal of Honor (1968)

    Languages -   German   Fluent
                     Spanish   Fluent
                     Chinese/Mandarin   Non-Fluent
                     Russian   Non-Fluent
                     Tibetan   Non-Fluent

    U.S. Passport Number: A 3073356 Issued December 24, 1980

References

    "Who's Who In Government" 1975 - 1976
    Upon request.

Science   Dr. Arthur L. Schawlow
J.T. Jackson - C.J. Wood Professor of Physics
Department of Physics
Stanford University
Palo Alto, California 94305
Nobel Laureate

Military   E.B. Roberts
Johnson & Anton Incorporated
3 Embarcadero, Suite 2020
San Francisco, California 94111
Lt. General, US Army, retired.

Religious Father Joseph Fessio
Director
St. Ignatius Institute
University of San Francisco
San Francisco, California 94117
Jesuit Order

Political Dr. Ray S. Cline
Senior Associate
Center for Strategic and
International Studies,
Georgetown University
1800 K Street NW, Suite 460
Washington, DC 20006
Former Deputy Director CIA and Chief INR,
US Department of State

Political Harold B. Chipman
President
Chipman Group Limited
550 Battery Street
San Francisco, California 94111
Merchant banking and risk analysis

Economic Donald K. Haenlein
President
Sarfatti Corporation
1000 Chestnut Street
San Francisco, California 94109
Entrepreneur

Economic Archibald B. Roosevelt, Jr.
Director - International Relations
The Chase Manhattan Bank, N.A.
1 Chase Manhattan Plaza
New York, New York 10015
Banking

Economic   Chester W. Bunnell, Jr.
            Director
            Henry F. Swift & Company
            433 California Street, Suite 800
            San Francisco, California 94104
            Stockbroker

Economic   Albert J. Morris
            Senior Vice President
            Bank of Californaia
            400 California Street
            San Francisco, California 94104
            Corporate Banking

Economic   John J. Altenberg
            Attorney at Law
            Executive Life Building
            9777 Wilshire Boulevard, Suite 607
            Beverly Hills, California 90212-1963
           Corporate-Legal

Economic Fortune 100 Business Executive - Open

Intellectual - Open

Social - Open

IV THE MARKET

Should the tests yield positive results, an immediate
market exists for pair-correlated photon generators used in
the test, but no specific market research for this device has
been conducted or required now. No organization presently
manufactures such devices. These devices would be
miniaturized for commercial use.

The Dr. Sarfatti pair-correlated photon switching
devices are capable of operating faster and at greater
capacities than used in the current state-of-the-art
computers and communications systems.

Specific applications include:

Computer Hardware - The devices would eliminate the
present requirement for sub-miniaturization and super
cooling. The devices would provide elements capable of
operating faster than currently used silicon chips, gallium

arsenide and the Josephson superconducting junctions under
development in "Fifth Generation Computers." The use of he
new switching devices would allow the construction of "Sixth
Generation Computers."

The construction of electron spin molecular computers
would be possible.

Secure communications for banking and diplomatic traffic
could be provided.

Military - Secure communications (unjammable and
untappable) could be provided the military in
command-control-communications for submarines, early warning,
NASA deep space probes and in key elements of the Strategic
Defense Initiative now under consideration in Washington.
Certain defensive weapons systems that would obsolete the
need for nuclear weaponry.

Other - Improvements could be made in spin resonance medical
diagnostic instruments, in molecular design of
pharmaceuticals and in genetic engineering. New quantum
telescopes could be developed, as well as potentially new
superphysics energy sources (zero-point vacuum phase
transitions).'

Competition

There is no known competition. Since the mathematical
calculations were completed only recently and are being held
in a safe desposit vault in a bank in California, no
theoretical physicist is aware of the completed version.
While it is possible that another qualified physicist could
have come to the same conclusion independently of the
inventor, the likelihood of this happening is remote. There
is no evidence available that a theoretical physicist or a
laboratory in the American private or public sector are
currently conducting tests in this area.

Because of the potential existing in the commercial and
military areas if the formula is proved correct, it is
possible that physicists in the Soviet Union are working on
the theory and its experimental tests. A member of the USSR
Academy of Sciences, Dr. Igor Akchurin, physicist and
Director of the Moscow Institute of Philosophy has shown
interest in Dr. Sarfatti's research since 1977.

In any foreign government where funds and talent are
available for pure research and where significant financial
and strategic payoffs are possible, it can be expected that
the foreign government would expedite testing Dr. Sarfatti's
new theory.

EMBASSY OF THE
UNITED STATES OF AMERICA
Paris, France

May 7, 1982

Mr. Jack Sarfatti
PO 26548
San Francisco, Calif.

Dear Jack,

Just a quick note to thank you for
your long letter and the "dossier" for TV
presentation. I am planning to see Alain
Aspect after the Reagan visit here and will
pass on to him whatever information you wish.
I would like to listen carefully to the tape
that Lawry and I made of our visit with
Mr. Aspect before attempting to describe
in what way he feels that Gary Zukav
misrepresented him. He does not consider his
experiment a test of faster than light
phenomena, no matter what the result. He was
on his way to a conference in Italy where,
I believe, they were going to discuss
experiments such as his which would also
include faster than light interpretation.

I hope I will have time in mid-June to
concentrate a little bit on all of this.
I am still settling in here as Lawry will tell
you. I am in constant touch with Guy Obolensky
who thinks that you and he are working on the
same wave length, the difference being that
Guy has made his discoveries through his
inventions and he is able, he believes, to
send signals (information) faster than light
and can demonstrate this phenomenon (once
he gets about $150,000 to set up a demonstration
properly). I have sent him a copy of your letter
to me as well as the "dossier". His address is:

124 Rte 17
Sloatsburg, N.Y. 10974

and his telephone number: (914) 753-2781

Jack Sarfatti

APENDIX B
DOD No. 86.1

U.S. DEPARTMENT OF DEFENSE
SMALL BUSINESS INNOVATION RESEARCH PROGRAM
Phase 1-FY 1986
PROJECT SUMMARY

Topic No. 77   United States Air Force

Name and Address of Proposing Small Business Firm:
Digitech Computer Services
1230 Grant Ave. #205
San Francisco, California 94133

Name and Title of Principal Investigator:
Dr. Jack Sarfatti
Director of Research

Title Proposed by Small Business Firm:
Research in the theoretical physics of zero-point quantum dynamic
energy of the metastable vacuum state.

Technical Abstract:
Theoretical Study of Quantum Vacuum Instability for Rocket Propulsion
The Forward report [AFRPL TR-83-067] only considered the
electromagnetic force. We propose to study the effect of the weak and strong
forces on vacuum stability in ultrastrong external electromagnetic fields. OF
particular interest is the Higgs mechanism for spontaneous broken symmetry of
the vacuum state of the weak force. The hope is that we can induce first order
phase transitions between metastable vacuum states that will release locked in
zero point fluctuation energy for propulsion. The big bang creation of the
universe was such an event. The question is whether we can control such a
process on a small scale. We will also consider the possibility of stimulated
beta decay at the quark level using the self interaction of the weak component
of a gamma ray beam resonant with the 3 mev down-up quark dynamic mass
difference. Though not directly related to propulsion, the latter might lead to
a new sort of beam weapon to render nuclear weapons "impotent and obsolete".
Finally, we shall study the role of supersymmetric fermion-boson transitions on
vacuum instability.
Ref. Sarfatti J., 1986, "Can the electroweak unified force be used to
neutralize nuclear weapons?", Defense Analysis, Vol. 2, No. 1 [in press]
Brasseys Defense Publishers, London-Washington-Toronto-Sydney-Paris-Frankfurt

Anticipated Benefits/Potential Commercial Applications of the Research
and Development:
1] alternate propulsion source for military space craft.
2] new sort of particle beam weapon to render nuclear missiles
impotent and obsolete.

Key words:
vacuum instability, zero point energy, stimulated beta
decay

Wavelets and Wigner Phase Space Density
Notes by Jack Sarfatti

The Wigner phase space density $W_f(x,p)$ of a "signal" f is

$$W_f(x,p) \equiv \frac{1}{2\pi} \int_{-\infty}^{+\infty} dx' e^{\frac{ipx'}{\hbar}} f\left(x+\frac{x'}{2}\right) f*\left(x-\frac{x'}{2}\right)$$

$$\equiv \frac{1}{2\pi} \int_{-\infty}^{+\infty} dp' e^{-\frac{ipx'}{\hbar}} \tilde{f}*\left(p+\frac{p'}{2}\right) \tilde{f}\left(p-\frac{p'}{2}\right)$$

(1.1)

$$\iint_{-\infty \to +\infty} dx dp W_f(x,p) = \int_{-\infty}^{+\infty} |f(x)|^2 dx = \int_{-\infty}^{+\infty} |\tilde{f}(p)|^2 dp$$

(1.2)

(1.2) is a generalized Parseval conservation relation of basis independence. What is the signal f in (1.1)? I am now not interested in generality but in my macro-quantum vacuum model specifically. You can consider the raw signal as the bound pair quantum wave function of a single virtual electron-positron pair with coordinates in 4-dim globally flat spacetime of the "false micro-quantum vacuum" of completely incoherent unbound virtual electron-positron pairs. This is a kind of quasi-Yilmaz space that Hal Puthoff likes, but there is no collective emergent gravity in this false vacuum. Emergent gravity in Andre Sakharov's sense out of the globally flat gauge force and source pre-geometrodynamic micro-quantum foam requires a Bose-Einstein condensation, or macroscopic occupation of the single-virtual pair bound state $\psi_{e^+-e^-}(x_{cm}, x_{rel})$ where x_{cm} is the center of mass coordinate and x_{rel} is the relative coordinate between the virtual electron and the virtual positron. That is, let

$$x_{e^+} \equiv x_{cm} + \frac{x_{rel}}{2}$$

$$x_{e^-} \equiv x_{cm} - \frac{x_{rel}}{2}$$

(1.3)

The pair density matrix is

$$\rho_{e^+e^-}\left(x_{e^+}, x_{e^-} \mid x_{e^+}', x_{e^-}'\right) \equiv \psi_{e^+e^-}\left(x_{e^+}, x_{e^-}\right)^* \psi_{e^+e^-}\left(x_{e^+}', x_{e^-}'\right)$$

$$= \psi_{e^+e^-}\left(x_{cm}, x_{rel}\right)^* \psi_{e^+e^-}\left(x_{cm}', x_{rel}'\right) \tag{1.4}$$

The reduced pair density matrix in the center of mass coordinates is

$$\rho^1_{e^+e^- cm}\left(x_{cm}, x_{cm}'\right) \equiv \iint dx_{rel} dx_{rel}' \, \psi_{e^+e^-}\left(x_{cm}, x_{rel}\right)^* \psi_{e^+e^-}\left(x_{cm}', x_{rel}'\right)$$

$$\tag{1.5}$$

Bose-Einstein condensation ("ODLRO"[323]) in the center of mass coordinates
is

$$\rho^1_{e^+e^- cm}\left(x_{cm}, x_{cm}'\right) \rightarrow \langle N_s \rangle \psi^*_{e^+e^- cm}\left(x_{cm}\right) \psi_{e^+e^- cm}\left(x_{cm}'\right) + \rho^1_{e^+e^- n}\left(x_{cm}, x_{cm}'\right)$$

$$\tag{1.6}$$

The first term on RHS of (1.6) is the smooth coherent nonrandom macro-quantum "superfluid". The second term is the residual locally random micro-quantum normal fluid from Heisenberg uncertainty noise that in vacuum we associate with "zero point fluctuations".

$$\Psi_{e^+e^- s}\left(x_{cm}\right) \equiv \sqrt{\langle N_s \rangle} \psi_{e^+e^-}\left(x_{cm}\right)$$

$$\langle N_s \rangle \gg 1 \tag{1.7}$$

$\langle N_s \rangle$ is the "large macroscopic eigenvalue" mean number of virtual pairs in the same center of mass "mother wavelet". The local macro-quantum BEC order parameter, whose phase modulation gives Einstein's $g_{\mu\nu}(x)$, is a coherent superposition of different integer occupation numbers of the mother wavelet. The density matrix for the center of mass motion of the pair in (1.6) is a correlation function. Define

---

[323] "Gauge Invariance in the Theory of Superfluid Helium", Ph.D. dissertation, Jack Sarfatti, UCR (August, 1969) F.W. Cummings of "Jaynes & Cummings", advisor.

$$x \equiv \frac{x_{cm} + x_{cm}'}{2}$$

$$x' \equiv x_{cm} - x_{cm}'$$

$$x_{cm} = x + \frac{x'}{2}$$

$$x_{cm}' = x - \frac{x'}{2}$$

$$(1.8)$$

The superfluid part of the Wigner phase space density, in special relativity notation, which is OK for pre-gravity micro-quantum gauge force and source "false vacuum", is therefore,

$$W_{e^+e^-s}(x,p) \equiv \left(\frac{1}{2\pi}\right)^4 \int_{-\infty}^{+\infty} d^4x' e^{i\frac{px'}{\hbar}} \Psi_{e^+e^-s}\left(x+\frac{x'}{2}\right) * \Psi_{e^+e^-s}\left(x-\frac{x'}{2}\right)$$

$$(1.9)$$

Similarly, for the normal random noisy "zero point fluctuations",

$$W_{e^+e^-n}(x,p) \equiv \left(\frac{1}{2\pi}\right)^4 \int_{-\infty}^{+\infty} d^4x' e^{i\frac{px'}{\hbar}} \rho^1_{e^+e^-n}(x_{cm}, x_{cm}')$$

$$= \left(\frac{1}{2\pi}\right)^4 \int_{-\infty}^{+\infty} d^4x' e^{i\frac{px'}{\hbar}} \rho^1_{e^+e^-n}\left(x+\frac{x'}{2}, x-\frac{x'}{2}\right)$$

$$(1.10)$$

I now drop the cm subscript in the coordinate notation. Take as the basic ("mother") wavelet, the effective center of mass wave packet $\psi^*_{e^+e^-cm}(x)$ of a single virtual electron-positron pair. Define the wavelet basis

$$\psi^*_{e^+e^-j,k}(x) \equiv 2^{j/2} \psi^*_{e^+e^-cm}\left(2^j x - k\ell\right)$$

$$j,k = 0, \pm 1, \pm 2, \ldots$$

$$(1.11)$$

Define the self-observing macro-quantum vacuum integral wavelet transform as

$$\widehat{\Psi}_{e^+e^-s:j,k}(\ell) \equiv \left\{2^{j/2}\right\}^4 \int_{-\infty}^{+\infty} \Psi_{e^+e^-s}(x)\psi_{e^+e^-cm}*\left(2^j x - k\ell\right)d^4x$$

(1.12)

where ℓ is the Kleinert world crystal lattice spacing ~ Planck scale.

Shall we take a Wigner phase space density of a wavelet, or a wavelet transform of a Wigner phase space density? Both may be interesting. Let's first look at the latter. Start with (1.9)'s $W_{e^+e^-s}(x,k)$. What about

$$W_{e^+e^-s:jk}(x,p) \overset{?}{\equiv} W_{e^+e^-s}\left(2^j x - k\ell, p/2^j + \hbar/k\ell\right)$$

(1.13)

The self-measuring wavelet transform of the superfluid Wigner phase space density is then

$$\widehat{W}_{e^+e^-s:jk}(\ell) \overset{?}{\square} \iint d^4x\, d^4p\, W_{e^+e^-s}(x,p)*W_{e^+e^-s}\left(2^j x - k\ell, p/2^j + \hbar/k\ell\right)$$

(1.14)

The adaptive phase space window area is

$$\left[\frac{k\ell}{2^j} + \frac{x_c}{2^j} - \frac{\Delta_{W_s}}{2^j}, \frac{k\ell}{2^j} + \frac{x_c}{2^j} + \frac{\Delta_{W_s}}{2^j}\right] \times \left[\frac{2^j\hbar}{k\ell} + 2^j p_c - 2^j \Delta_{\tilde{W}_s}, \frac{2^j\hbar}{k\ell} + 2^j p_c + 2^j \Delta_{\tilde{W}_s}\right]$$

(1.15)

Where obviously

$$x_c \equiv \iint x W_{e^+e^-s}(x,p)\, d^4x\, d^4p$$

$$p_c \equiv \iint p W_{e^+e^-s}(x,p)\, d^4x\, d^4p$$

$$\Delta_{W_s}^2 \equiv \iint (x - x_c)^2\, W_{e^+e^-s}(x,p)\, d^4x\, d^4p$$

$$\Delta_{\tilde{W}_s}^2 \equiv \iint (p - p_c)^2\, W_{e^+e^-s}(x,p)\, d^4x\, d^4p$$

(1.16)

Zero Point Energy Gravity Physics
Of UFOs, Star Gates, Time Travel and Parallel Brane Worlds

A Pedagogical Introduction
By
Jack Sarfatti
Click on the equations to make them bigger.[324]

1. Newton's Classical Gravity Potential Theory in a Nutshell

The static near field[325] Green's function is

$$G\left(\vec{r}-\vec{r}\,'\right) \equiv \frac{-4\pi G}{\left|\vec{r}-\vec{r}\,'\right|}$$

$$(1.1)$$

where

$$\vec{\nabla}^2 G\left(\vec{r}-\vec{r}\,'\right) = -4\pi G\delta^3\left(\vec{r}-\vec{r}\,'\right)$$

$$(1.2)$$

---

[324] Like in Lewis Carroll's "Alice in Wonderland". ☺ Only in htm not pdf versions using MathType5. It may not work in MAC OS. Go to http://stardrive.org/math4/

[325] The dispersion relation (AKA "mass shell") $\omega \equiv 2\pi\nu = c\left|\vec{k}\right|$ for far field gravity waves is violated in the near field for frequency ν Hertz and wave 3-vector $\vec{k} \equiv \vec{p}/\hbar$. The entire $\omega - \vec{k}$ 4-space contributes to dynamic near fields, unlike the case of far field radiation. The near field is a macro-quantum coherent state of "virtual quanta" of the boson field. All Bose-Einstein condensates (BEC) are macro-quantum coherent, but not vice versa.

The universal[326] attractive static gravitational potential energy $U(\vec{r})$ per unit test particle of mass m at the field point with displacement 3-vector $\vec{r}$ [327] for source mass density $\rho(\vec{r}')$ at source point $\vec{r}'$ is[328]

$$U(\vec{r}) = \iiint G(\vec{r} - \vec{r}')\rho(\vec{r}')d^3r'$$

$$= -4\pi G \iiint \frac{\rho(\vec{r}')}{|\vec{r} - \vec{r}'|}d^3r'$$

(1.3)

The Newtonian gravitational local Poisson equation precursor to Einstein's 1915 geometrodynamic field equation is

$$\vec{\nabla}_r^2 U(\vec{r}) = \iiint \left[\vec{\nabla}_r^2 G(\vec{r} - \vec{r}')\right]\rho(\vec{r}')d^3r'$$

$$= -4\pi G \iiint \delta^3(\vec{r} - \vec{r}')\rho(\vec{r}')d^3r'$$

$$= -4\pi G \rho(\vec{r})$$

(1.4)

The universal force per unit test particle[329] mass field, is therefore,

---

[326] Universality incorporates the Galilean principle of equivalence called the weak principle of equivalence in Einstein's general theory of relativity (GR). That is, all bodies fall with the same acceleration in the same place in a gravitational field.

[327] Relative to an arbitrary origin of coordinates in a global Galilean frame of reference.

[328] Note that the dimensions of both sides of equation (1.3) are speed squared. The dimensions of Newton's G are 1/(mass density) x (time)^2. This is useful to commit to memory.

[329] The test particle approximation ignores direct back-reaction by the test particle on whatever is influencing it. In Bohm's realist quantum theory, the inability to clone a photon comes from treating the hidden variable as a test particle. That is, the qubit pilot field moves the hidden variable particle without the particle directly distorting the pilot field. That is the orthodox statistics of quantum theory with the linear evolution of the density matrix depends on this approximation of action without reaction. See p. 30 & Ch 14 of "The Undivided Universe" by David Bohm and Basil Hiley (Routledge, 1993) and Asher Peres

$$\vec{f}_g(\vec{r}) \equiv -\vec{\nabla}U(\vec{r}) = -4\pi G \iiint \frac{\rho(\vec{r}\,')(\vec{r}-\vec{r}\,')}{\left|\vec{r}-\vec{r}\,'\right|^3} d^3 r'$$

$$(1.5)$$

In the special case of an isolated point mass source

$$\rho(\vec{r}\,') \rightarrow M\delta^3(\vec{r}\,')$$

$$(1.6)$$

$$\vec{f}_g(\vec{r}) \rightarrow = -4\pi G \iiint \frac{M\delta^3(\vec{r}\,')(\vec{r}-\vec{r}\,')}{\left|\vec{r}-\vec{r}\,'\right|^3} d^3 r'$$

$$= -4\pi G \frac{M\vec{r}}{\left|\vec{r}\right|^3}$$

$$(1.7)$$

Note from equation (1.7) for a point source mass M, that taking the radial derivative gives a minus sign that is cancelled in the minus sign of the negative gradient. With this convention, one needs the $-$ sign in the static Green's function to get an attractive force back to the source for a positive mass density.

2. Einstein's General Theory of Relativity in the Galilean-Newtonian Limit

The local non-gravitational classical field[330] stress energy tensor $T_{\mu\nu}$ has the form of a general relativistic fluid

---

"How the No-Cloning Theorem Got its Name", http://xxx.lanl.gov/abs/quant-ph/0205076

[330] The Yilmaz theory, with a nonvanishing local pure classical vacuum gravity field stress-energy tensor, found implicitly in Hal Puthoff's "PV" theory, is rejected as a violation of the Einstein equivalence principle (EEP) as explained in detail by Misner, Thorne & Wheeler in "Gravitation". For this reason alone, as well as several others equally fatal, Puthoff's PV gravity claims are completely without any scientific foundation in my opinion. For the record, my remarks here

$$T_{\mu\nu} = g_{\mu\sigma} g_{\nu\varsigma} \left(\rho + \frac{p}{c^2} \right) \frac{dx^\sigma}{ds} \frac{dx^\varsigma}{ds} - p g_{\mu\nu}$$

(2.1)

Where ρ is the mass density and p is the pressure.[331] The trace[332] of this second rank tensor is an absolute objectively real scalar invariant, i.e. the same measurable real number for all observers in arbitrary motion relative to each other all measuring the same properties of the same events.

$$T \equiv g^{\mu\nu} T_{\mu\nu} = T_\mu^\mu = c^2 \rho - 3p$$

(2.2)

A key qualitative physical difference between Einstein's early 20th Century covering geometrodynamic theory of Newton's late 17th Century theory of gravity is that pressure is a local source of gravity in addition to mass density. The equation of state connects pressure to mass density. Einstein's local field equation, with zero cosmological Λ local field, can be written as

$$R_{\mu\nu} = -\frac{8\pi G}{c^4} \left(T_{\mu\nu} - \frac{1}{2} g_{\mu\nu} T \right)$$

(2.3)

The dominant term in the Newtonian limit is the time-time term i.e.

$$R_{00} = -\frac{8\pi G}{c^4} \left(T_{00} - \frac{1}{2} g_{00} T \right)$$

(2.4)

Where $g_{00} \to 1$, therefore

---

should *not* be over-extrapolated and misconstrued as a blanket condemnation of Puthoff's scientific work in other more practical engineering physics fields.

[331] Pressure is defined as the diagonal elements of the 3-dim space-space stress tensor piece of the 4-dim total local stress energy tensor of non-gravitational field, here assumed isotropic for simplicity.

[332] Sum of the 4 diagonal elements.

$$R_{00} \rightarrow -\frac{8\pi G}{c^4}\left(c^2\rho - \frac{1}{2}\left(c^2\rho - 3p\right)\right) = -\frac{4\pi G}{c^4}\left(c^2\rho + 3p\right) \approx \frac{1}{c^2}\vec{\nabla}^2 U$$

$$(2.5)$$

Therefore, Einstein's generalization of Newton's Poisson equation in the weak field limiting approximation of large principal radii of curvature compared to the scale of the system is

$$\vec{\nabla}^2 U = -4\pi G\left(\rho + \frac{3p}{c^2}\right)$$

$$(2.6)$$

A positive active source density $\rho + 3p/c^2$ giving an overall minus sign on the RHS of equation (2.6) is a universally attractive gravitating source region, similarly, an overall plus sign is a universally repelling anti-gravitating source region. Note that a dipole of opposite signs of active source density will self-accelerate as in "propellantless propulsion", or Paul Hill's "acceleration field."[333] This was first pointed out by Hermann Bondi in the 1950's when he was Chief Scientist of the British Ministry of Defense. I heard his lecture on this at Cornell when I was an undergraduate physics major in Hans Bethe's senior honors seminar in 1960. I was also studying tensor calculus with Wolfgang Rindler at the time and was Paul Olum's grader in the Cornell Math Department for engineering calculus. The full significance of Bondi's remark was not understood until today more than 40 years later.

All classical equations of state have a strictly positive active source density, i.e.,

$$\rho_{classical} + \frac{3p_{classical}}{c^2} > 0$$

$$(2.7)$$

---

[333] "Unconventional Flying Objects" by Paul Hill who was a solid sober USG aeronautical engineer circa post WWII at time of Roswell. His book is the best technical engineering book on UFOs, indeed the only one that is not crackpot.

Indeed, this is a crucial assumption in the Penrose-Hawking classical spacetime singularity theorems based on global topology independent of details of the local geometrodynamical field equation and the classical action it comes from. Hal Puthoff is not cognizant of this when he claims that his ill-posed classical PV theory has no black holes. Quantum physics changes this. Indeed, Heisenberg's uncertainty principle combined with Einstein's local principle of equivalence (EEP) implies[334] the following universal zero point energy equation of state for the vacua of all micro-quantum fields both boson and fermion

$$\rho_{vac} = -\frac{p_{vac}}{c^2}$$
(2.8)

Therefore, in general, the active micro-quantum vacuum field source density is

$$\rho_{vac} + \frac{3p_{vac}}{c^2} = -2\rho_{vac}$$
(2.9)

Field	Sign of ρ_{vac}	Vacuum gravity
Boson	Positive	Repulsive
Fermion	Negative	Attractive

The micro-quantum vacuum Poisson equation is then

$$\vec{\nabla}^2 U_{vac} = +8\pi G \rho_{vac}$$
(2.10)

The local stress-energy tensor of this quantum vacuum in terms of the local cosmological $\Lambda(x)$ is

---

[334] Eq. (1.88) p. 26 John Peacock, "Cosmological Physics", Cambridge Press ,(the Cal Tech text book).

$$t_{\mu\nu(vac)} = \frac{\Lambda c^4}{8\pi G} g_{\mu\nu}$$

$$(2.11)$$

Therefore,

$$t_{00vac} \equiv \rho_{vac}c^2 = \frac{\Lambda c^4}{8\pi G} g_{00} = \frac{\Lambda c^4}{8\pi G}$$

$$(2.12)$$

Substitute (2.12) into (2.10)

$$\vec{\nabla}^2 U_{vac} = c^2\Lambda$$

$$(2.13)$$

This is our key equation in the weak field limit, which illustrates the key new insight on how advanced super-technology will work.

$\Lambda > 0$ vacuum regions correspond to Kip Thorne's anti-gravitating "exotic matter" needed to make Star Gates and Time Machines. At the cosmological scale they explain the observed acceleration of our visible universe. Similarly $\Lambda < 0$ vacuum regions correspond to the "dark matter" that is at least ~ 80% of the total effective gravitating mass of our universe. A photon leaving a $\Lambda > 0$ region detected in a region will be blue shifted. Similarly a photon leaving a $\Lambda < 0$ region detected in a region will be red shifted. Hawking's equations for the holographic universe are modified as well.[335]

3. Hawking's Holographic Universe Equations

From: Jack Sarfatti [sarfatti@pacbell.net]
Sent: Tuesday, June 11, 2002 11:37 AM

---

[335] Alleged alternating red and blue shifts from UFOs have been written about, e.g. Eric Davis's 2001 MUFON paper
http://198.63.56.18/pdf/davis_mufon2001.pdf

To: sarfatti@well.com; Jack Sarfatti
Cc: ItalianPhysicsCenter
Subject: Hawking's Nutty Universe: UFO Beauties from The Future?

From "The Universe in a Nutshell" Ch V

"So if a beautiful alien in a flying saucer invites you into her time machine. Step with care."

No, that's not from Nick Herbert's "Quantum Tantrum"
http://members.cruzio.com/~quanta/

It's not from

http://stardrive.org/cartoon/coffee.html

It's on p. 144 of Hawking's new book.

Also one finds:

"It's tricky to speculate openly about time travel. One risks either an outcry at the waste of public money... or a demand that the research be classified for military purposes. ...There are only a few of us foolhardy enough to work on a subject that is so politically incorrect in physics circles. We disguise the fact by using technical terms that are code for time travel (p. 133) ... we have no reliable evidence of visitors from the future. I'm discounting the conspiracy theory that UFOs are from the future and that the government knows and is covering up. Its record of cover-ups is not that good (p. 142).... You might wonder if this chapter is part of a government cover-up on time travel. You might be right. (p. 153)"

Thus, Hawking sets the stage perfectly for my book "Destiny Matrix"

http://stardrive.org/Jack/cover.jpg

See also "The Star Gate Conspiracy" by Picknett & Prince.

http://www.templarlodge.com/

http://www.cassiopaea.org/perseus/bearden.htm

Previously Jack wrote:

Hawking gives two basic formulae for the world hologram entropy S and temperature T associated with classical curved spacetime that is effectively 3D rather than 4D (Ref: "The Universe in a Nutshell").

$$S = \frac{Akc^3}{4\hbar G} = \frac{A}{4L_p^2}k \tag{3.1}$$

For our holographic universe pp 64-65 Hawking and

$$T = \frac{\hbar c^3}{8\pi kGM} = \frac{\hbar c}{4\pi k \frac{2GM}{c^2}} = \frac{\hbar c}{4\pi kR_s} = \frac{1}{4\pi k}\left(\frac{\hbar}{Mc}\right)\frac{c^4}{G} \tag{3.2}$$

where for a nonrotating black hole

$$A \rightarrow 4\pi\left(\frac{2GM}{c^2}\right)^2 \tag{3.3}$$

k = Boltzmann's "imaginary time" quantum of entropy (disorder) ~ 1.4×10^{-16} ergs per degree Kelvin

$\hbar$ = Planck's "real time" quantum of classical mechanical action or periodic quantum phase ~ 10^{-27} erg-seconds

c = speed of far field transverse polarized electromagnetic waves (real photons on light cone "mass shell") ~ 3×10^{10} cm per second

G = Newton's constant of gravity ~ $6.7 \ 10^{-8}$ cm^3 per second^2 per gram = 6.7×10^{-8} per mass density-second^2

L_p^2 is the quantum of area ~ 10^{-66} cm^2. It is one quantum gravity c-bit of Shannon information of the World Hologram.

$R_s \equiv 2GM/c^2$ is the classical gravity radius for the event horizon[336] of a non-rotating black hole of mass M

$\hbar/Mc$ is the quantum "Compton wavelength" of the mass M. If you probe a region inside this distance you create real particles and anti-particles of mass M.

c^4/G is the superstring tension also called the reciprocal of the "spacetime stiffness factor" because Einstein's local geometrodynamical field equation is

$$G_{\mu\nu} + \Lambda g_{\mu\nu} = -\frac{G}{c^4}T_{\mu\nu}$$

(3.4)

The cosmological "constant" Λ (not really constant) has dimensions of 1/Area.

Up until recently[337] observation set a limit

$$|\Lambda| < 3\times10^{-56}\,cm^{-2} \sim \frac{3}{\left(10^{28}\,cm\right)^2} \sim 3\left(\frac{H}{c}\right)^2$$

(3.5)

Where H is the Hubble constant. Therefore $1/\Lambda$ is the World Hologram "area" $A_{universe} \sim 10^{122}$ bits.

The Wigner phase space density W(x,p) http://stardrive.org/math2/Wigner.htm at "wavelet" scale factor p can have sub-Planckian structure as shown by W. Zurek in Nature, Aug 2001.

Think of adding resistances in a parallel circuit

---

[336] Where time stops for external observer.
[337] Before accelerating universe data.

$$\frac{1}{A} = \frac{1}{4\pi}\left(\frac{c^2}{2GM}\right)^2 + \Lambda$$

(3.6)

or

$$A(\Lambda) = \frac{1}{\left(\dfrac{c^2}{8\pi GM}\right)^2 + \Lambda} = \frac{\left(\dfrac{8\pi GM}{c^2}\right)^2}{1 + \left(\dfrac{8\pi GM}{c^2}\right)^2 \cdot \Lambda} \equiv \frac{A_{Hawking}}{1 + A_{Hawking}\Lambda}$$

(3.7)

Dark matter is $\Lambda < 0$ corresponds to macro-quantum weird gravitating vacuum where the normal fluid Wigner phase space density is negative from giant superfluid macro-quantum interference in the virtual fermion-antifermion bound state local order parameter of spontaneous broken symmetry from false 100% normal fluid high entropy micro-quantum random vacuum to two fluid macro-quantum lower entropy true vacuum.

Hawking's anti-gravitating "vacuum energy" is $\Lambda > 0$, e.g. "the effect of vacuum energy is the opposite of that of matter ... vacuum energy causes the expansion to accelerate" p. 96. Note the sign dependence. Anti-gravitating $\Lambda > 0$ decreases the effective mass $M_{effective}$ and it also decreases the effective holographic entropy. Gravitating $\Lambda < 0$ dark energy does just the opposite.

$$\left(\frac{c^2}{2GM_{effective}}\right)^2 \equiv \left(\frac{c^2}{2GM}\right)^2 + 4\pi\Lambda$$

(3.8)

Therefore

$$M_{effective}(\Lambda) = \frac{M}{\sqrt{1+4\pi\left(\dfrac{2GM}{c^2}\right)^2 \Lambda}} = \frac{M}{\sqrt{1+A_{Hawking}\Lambda}}$$

(3.9)

Is the quintessent Λ field correction to the effective mass.

Similarly for holographic Hawking entropy and absolute temperature of the horizons and holographic areas in general.

$$S(M,\Lambda) = \frac{S_{Hawking}(M)}{1+4\pi\left(\dfrac{2GM}{c^2}\right)^2 \Lambda} = \frac{S_{Hawking}(M)}{1+A_{Hawking}\Lambda}$$

(3.10)

$$T(M,\Lambda) = \frac{\hbar c^3}{8\pi k GM_{effective}} = T_{Hawking}(M)\sqrt{1+A_{Hawking}\Lambda}$$

(3.11)

These are my quintessent Λ field corrections, or renormalizations, for quantum gravity thermodynamics started by Bekenstein and further developed by Hawking & Co

$\Lambda > 0$ decreases black hole entropy and increases black hole temperature.

$\Lambda < 0$ increases black hole entropy and decreases black hole temperature.

This generalizes to ordinary curved spacetime regions in the sense of the "world hologram" idea of Lenny Susskind[338] et-al.

---

[338] I worked with Lenny Susskind at Cornell in 1963 where we were both grad students. I had brought "space kid" Johnny Glogower to Cornell with Phil Morrison's help. All three of us worked on the quantum phase-time operator problem that I had started Lenny on. The problem was first suggested to me by George Parrent, a student of Emil Wolf's, when I worked at Tech/Ops in Burlington, Mass between Brandeis and going back to Cornell. The problem is mentioned in my Nuovo Cimento paper on correlations in black body radiation that I wrote at Tech/Ops. Lenny went ahead, at Peter Carruthers' suggestion, and

BTW check my algebra above in case I made stupid errors. ;-)

This is the first time I have seen these relationships cosmic triggered by reading Hawking's popular book.

There are two senses of "hologram" in Hawking's book. The first one is above in which information on three-dimensional space is coded on the two-dimensional bounding surface. The second meaning is that of a four-dimensional boundary of a five-dimensional space. I am not clear on how these two ideas connect.

"Information about the quantum states in a region of spacetime may be somehow coded on the boundary of the region which has two dimensions less. This is like the way that a hologram carries a three-dimensional image on a two-dimensional surface. If quantum gravity incorporates the holographic principle, it may mean that we can keep track of what is inside black holes.... If we can't do that, we won't be able to predict the future as fully as we thought."

The second meaning is:

"we may live on a 3-brane – a four dimensional (three space plus one time) surface that is the boundary of a five-dimensional region, with the remaining dimensions curled up very small. The state of the world on a brane encodes what is happening in the five-dimensional region." – Hawking p. 64 "The Universe in a Nutshell".

I somewhat disagree with Hawking's statement on p. 124 where Hawking says that faster than light communication by quantum nonlocality is

---

published a paper in Physics in the same volume the famous Bell inequality paper, without putting my name on it. I was at Ford Philco Aeronutronics in Newport Beach, CA at the time with Fred W. Cummings who later became my Ph.D. dissertation advisor at UCR (August, 1969) though all my class work was done at UCSD in La Jolla. Indeed, Fred liked to come down to La Jolla to work at the Scripps Oceanographic Institute on the ocean.

"ridiculous". True, it cannot happen in orthodox quantum theory and Tony Valentini shows why in a way that shows a loop hole.

http://users.ox.ac.uk/~quee0776/valentiniabs.html

http://www.fourmilab.ch/rpkp/valentini.html

http://www.edge.org/3rd_culture/bios/valentini.html

http://www.edge.org/discourse/information.html

Hawking keeps talking about the need for negative energy density. We do not really need it when the quintessent field $\Lambda > 0$ in the macro-quantum vacuum in limited regions. Light diverges there just like Hawking needs.

The world hologram idea requires a coherent phase field. This is indeed my macro-quantum vacuum coherent complex numbered order parameter $\Psi(x)$.

$$\Psi(x) \equiv \sqrt{\frac{1}{L_p^3}\left(1 - L_p^2 \Lambda(x)\right)} \sum_P e^{i\left[\arg \Psi(x) - \frac{2e}{\hbar c}\int_P^x A_\mu(x')dx'^\mu\right]}$$

$$= \sqrt{\frac{1}{L_p^3}\left(1 - L_p^2 \Lambda(x)\right)} e^{i\arg \Psi(x)} \sum_P e^{-i\frac{2e}{\hbar c}\int_P^x A_\mu(x')dx'^\mu}$$

$$\equiv \sqrt{\rho_s(x)} e^{i\arg \Psi(x)} \tag{3.12}$$

The sum $\overset{\Sigma}{P}$ is over all Feynman paths P in the false globally flat vacuum "Flat World"[339] that terminate at event x. Define, the non-integrable renormalization factor[340]

---

[339] Flat World has the Fermi surface without bound state pairing of the virtual electron-positron pairs to a lower more stable energy. It is 100% "Dirac sea".

[340] $|Z(x)|$ can be absorbed into the Planck scale $L_p \to L_p^*(x)$ making is larger and variable. The phase $\arg Z$ has singularities (e.g. vortex lines) representing

$$\sum_P e^{-i\frac{2e}{\hbar c}\int_P^x A_\mu(x')dx'^\mu} \equiv \left|Z(x)\right| e^{i\arg Z(x)}$$

$$(3.13)$$

$$\rho_s(x) \equiv \frac{1}{L_p^3}\left(1 - L_p^2 \Lambda(x)\right)\left|\sum_P e^{-i\frac{2e}{\hbar c}\int_P^x A_\mu(x')dx'^\mu}\right|^2$$

$$= \frac{1}{L_p^3}\left(1 - L_p^2 \Lambda(x)\right)\left|Z(x)\right|^2$$

$$\equiv \rho(x) - \rho_n(x)$$

$$(3.14)$$

$$\rho(x) \equiv \frac{\left|Z(x)\right|^2}{L_p^3}$$

$$\rho_n(x) \equiv \frac{\Lambda(x)\left|Z(x)\right|^2}{L_p}$$

$$(3.15)$$

$Z(x)$ is the electromagnetic control renormalization factor for the macro-quantum vacuum of Curve World that comes from the spontaneous breakdown of the global symmetry of micro-quantum Flat World. This is analogous to my 1966 UKAERE paper with Marshall Stoneham on the "Goldstone Theorem and the Jahn-Teller Effect". Note the Bohm-Aharonov-Josephson phase integral in the coherent sum over both *future* and past Feynman *destinies* and histories, respectively ending on event x in the "here-now".

---

non-integrable hysteresis memories from both past *and future* (e.g. Yakir Aharonov et-al "Destiny and History quantum state vectors").

$$\rho(x) \equiv \frac{|Z(x)|^2}{L_p^3} \equiv \frac{1}{L_p^{*3}(x)}$$

$$\rho_n(x) \equiv \frac{\Lambda(x)|Z(x)|^2}{L_p} \equiv \frac{\Lambda^*(x)}{L_p^*(x)}$$

$$L_p^*(x) \equiv \frac{L_p}{|Z(x)|^{2/3}}$$

$$\Lambda^*(x) \equiv \Lambda(x)|Z(x)|^{4/3} \tag{3.16}$$

For now I only include the electromagnetic field and the virtual electron-positron pairs that Bose-Einstein condense into the same cell in phase space of volume $\sim h^3$ in the center of mass coordinate. The relative coordinates are integrated out. I then form the *virtual superfluid* Wigner phase space density[341] of the macro-quantum vacuum substrate of the classical Einstein world from this nonrandom *zero entropy* coherent order parameter[342]

$$W_s(x,p) \equiv \left(\frac{1}{h}\right)^4 \int e^{ipy/\hbar} \Psi^* \left(x - \frac{y}{2}\right) \Psi \left(x + \frac{y}{2}\right) d^4y \tag{3.17}$$

There is also a corresponding virtual incoherent random "zero-point fluctuation" normal fluid component that carries all the entropy. The virtual superfluid component of the physical vacuum's Wigner phase space density has zero entropy which explains the mystery of the arrow of time.

$$W_{ZPF}(x,p) \equiv \left(\frac{1}{h}\right)^4 \int e^{ipy/\hbar} \rho_{ZPF} \left(x - \frac{y}{2}, x + \frac{y}{2}\right) d^4y \tag{3.18}$$

---

[341] The Wigner phase space density can go negative. Indeed, the vacuum superfluid Wigner density is negative in huge regions of "dark matter" in the universe.

[342] "Sub-Planck structure in phase space and its relevance for quantum decoherence", W. H. Zurek, Nature, 412, p. 712, 16 August, 2001.

Negative regions of the Wigner phase space density of the "normal fluid" random virtual zero point fluctuations are the gravitating dark matter regions of $\Lambda^*(x) < 0$.

$$W_s(x,p) + W_{ZPF}(x,p) = \frac{1}{\hbar^4}$$

(3.19)

We are free to choose complementary window 4-cell widths $\Delta^4 x, \Delta^4 p$ with x, p as center points in a wavelet analysis. The window cell sizes $\Delta^4 x$ in ordinary curved spacetime emerging from the long-range macro-quantum phase coherence must be small compared to the four local principal radii of curvature in a neighborhood of x.

$$\rho_s\left(x,p\mid\Delta_p^4\right) \equiv \rho_s\left(x,p;x,p\mid\Delta_p^4\right) \equiv W_s(x,p)\Delta_p^4$$

$$\rho_n\left(x,p\mid\Delta_p^4\right) \equiv W_n(x,p)\Delta_p^4$$

$$\Psi\left(x,p\mid\Delta_p^4\right) \equiv \sqrt{\rho_s\left(x,p\mid\Delta_p^4\right)}\,e^{i\Theta\left(x,p\mid\Delta_p^4\right)}$$

$$\rho_s\left(x,p;x'p'\mid\Delta_p^4\right) \equiv W_s(x,p)\Delta_p^4 e^{i\left[\Theta\left(x,p\mid\Delta_p^4\right)-\Theta\left(x',p'\mid\Delta_p^4\right)\right]}$$

$$\Theta\left(x,p\mid\Delta_p^4\right) \equiv \arg\Psi\left(x,p\mid\Delta_p^4\right) - \arg Z\left(x,p\mid\Delta_p^4\right)$$

(3.20)

$$\rho\left(x,p\mid\Delta_p^4\right) \equiv \frac{\left|Z\left(x,p\mid\Delta_p^4\right)\right|^2}{L_p^3}$$

$$\rho_n\left(x,p\mid\Delta_p^4\right) \equiv \frac{\Lambda\left(x,p\mid\Delta_p^4\right)\left|Z\left(x,p\mid\Delta_p^4\right)\right|^2}{L_p}$$

$$\rho_s\left(x,p\mid\Delta_p^4\right) \equiv \rho\left(x,p\mid\Delta_p^4\right) - \rho_n\left(x,p\mid\Delta_p^4\right)$$

(3.21)

$$\rho\left(x, p \mid \Delta_p^4\right) \equiv \frac{\left|Z\left(x, p \mid \Delta_p^4\right)\right|^2}{L_p^3} \equiv \frac{1}{L_p^{*3}\left(x, p \mid \Delta_p^4\right)}$$

$$\rho_n\left(x, p \mid \Delta_p^4\right) \equiv \frac{\Lambda\left(x, p \mid \Delta_p^4\right)\left|Z\left(x, p \mid \Delta_p^4\right)\right|^2}{L_p} \equiv \frac{\Lambda^*\left(x, p \mid \Delta_p^4\right)}{L_p^*\left(x, p \mid \Delta_p^4\right)}$$

$$L_p^*\left(x, p \mid \Delta_p^4\right) \equiv \frac{L_p}{\left|Z\left(x, p \mid \Delta_p^4\right)\right|^{2/3}}$$

$$\Lambda^*\left(x, p \mid \Delta_p^4\right) \equiv \Lambda\left(x, p \mid \Delta_p^4\right)\left|Z\left(x, p \mid \Delta_p^4\right)\right|^{4/3}$$

(3.22)

Einstein's *scale-dependent* classical curved spacetime metric tensor is

$$g_{\mu\nu}\left(x, p \mid \Delta_p^4\right) \equiv \frac{1}{2}\left\{\partial_\mu, \partial_\nu\right\}\left[L_P^{*2}\left(x, p \mid \Delta_p^4\right)\Theta\left(x, p \mid \Delta_p^4\right)\right]$$

(3.23)

Where { , } is the anti-commutator. Note that I have derived $L_p^*\left(x, p \mid \Delta_p^4\right)$ of Abdus Salam's "f-strong-short range gravity" model[343] from the 1970's that was the precursor for modern ideas of strong gravity from extra hyperspace dimensions.

"Large extra dimensions are an exciting new development ... They would imply that we live in a brane world, a four-dimensional surface or brane in a higher dimensional spacetime. Matter and nongravitational forces like the electric force would be confined to the brane ... On the other hand, gravity in the form of curved space would permeate the whole bulk of the higher dimensional spacetime .... Because gravity would spread out in the extra dimensions, it would fall off more rapidly with distance than one would expect ... If this more rapid

---

[343] I worked with Abdus Salam at ICTP in Trieste, Italy in 1973 on these ideas. See my papers in "Collective Phenomena" edited by Herbert Frohlich and my UCR Ph.D. dissertation advisor, Fred W. Cummings, (Gordon & Breach ?) in that period. Also my 1966 paper with Marshall Stoneham of Harwell UKAERE on spontaneous broken symmetry in solid state "Jahn-Teller Effect" in Proceedings of the Physical Society of London cited in American Institute of Physics "Resource Letter on Symmetry in Physics".

falloff of the gravitational force extended to astronomical distances, we would have noticed its effect on the orbits of the planets ... they would be unstable... However, this would not happen if the extra dimensions ended on another brane not that far away from the brane on which we live. Then for distances greater than the separation of the branes, gravity would not be able to spread out freely but would be confined to the brane, like the electrical forces, and fall off at the right rate for planetary orbits."

- Stephen Hawking, Ch. 7

4. The Macro-Quantum Vacuum

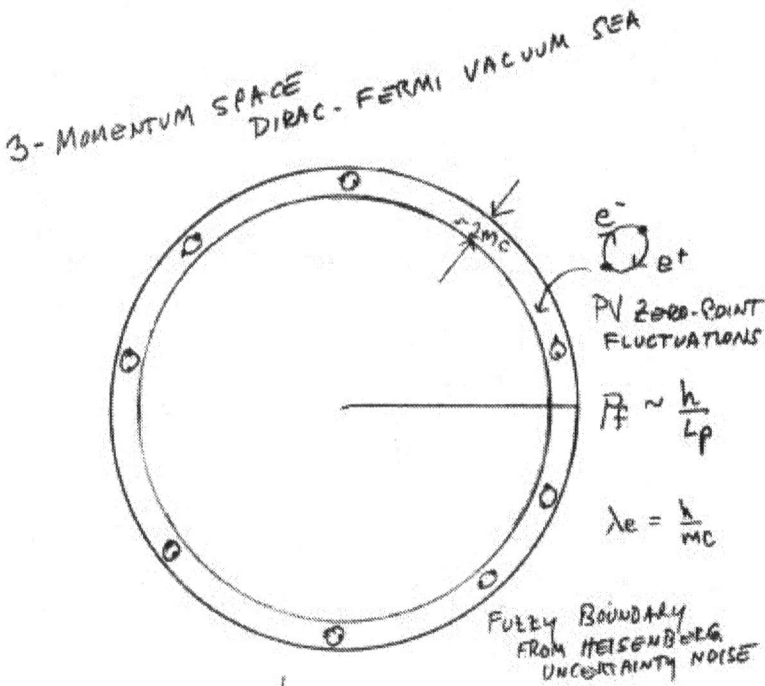

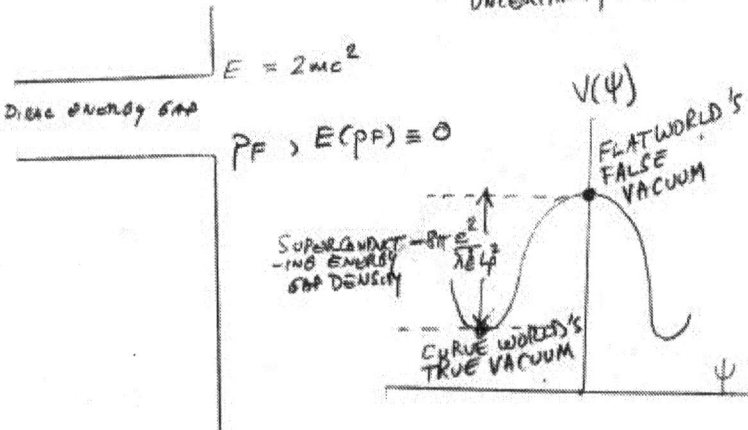

The false micro-quantum vacuum of "Flat World" is from gauge source and force fields. There is no gravity as yet which is, as Andre Sakharov first suggested, an emergent macro-quantum collective mode - indeed Einstein's 1915 "geometrodynamic" local field theory of gravity is a "signal modulation" of the stable long range coherent phase of a virtual Bose-Einstein condensate of a huge number of gauge source fermion-antifermion pairs whose centers of mass macroscopically occupy the same "mother wavelet" bound state. Special relativity works globally in the false vacuum of Flat World.

Think of the false vacuum of Flat World as a ball, i.e. a Fermi sphere in momentum space, of virtual electrons closely packed into negative energy states, i.e. the "Dirac Sea" as required by the Pauli exclusion principle. There are also virtual photons but they play a minor role for now. To make life simpler in this broken (symmetry) toy model of the world, let's suppose there is only quantum electrodynamics. We can worry about the complications of weak and strong forces later. The edge of this Fermi sphere is not sharp because of Heisenberg's uncertainty principle. Indeed the edge is fuzzed out with locally uncontrollably random quantum noise by a thickness of momentum $\sim 2mc$ below the threshold for real electron-positron pair creation, where m is the rest mass of the electron. The edge of the Fermi sphere has momentum $pf \sim h/Lp$, where $Lp^2 \sim hG/c^3 \sim 10^{-66}$ $cm^2 \sim 1$ Bekenstein-Hawking bit of entropy. Therefore, what we have is a fuzzy noisy spherical shell of phase space volume $\sim 4\pi(h/Lp)^2(2mc)V = 4\pi\ h^2(c^3/hG)(2mc)V = 4\pi h(c^3/G)(2mc)V = 8\ \pi hm(c^4/G)V$. There is an addition spin factor $2S + 1 = 2$. So the phase space volume of the fuzzy Heisenberg edge of the Fermi-Dirac sea is $\sim 16\pi hm(c^4/G)V$. Note that c^4/G is the super string tension of 10^{19} Gev per 10^{-33} cm (AKA reciprocal spacetime stiffness factor). Each "single-particle"[344] cell of phase space has volume h^3 therefore the false vacuum density of Planck action cells of phase space is $16\pi(mc/h)(c^3/hG)$ per unit ordinary space volume. The attractive Coulomb bound state energy of the virtual electron-positron (hole) pair is $\sim -e^2/(h/2mc)$. Therefore, the total "superconducting" energy density gap is:

Superconducting macro quantum vacuum "BCS" energy gap density

---

[344] The center of mass of a bound pair, real or virtual, acts like a single particle.

$$\sim - 8\pi (mc/h)^2 (c^3/hG)e^2$$

$$= - 8\pi e^2/(\text{Compton wave length of electron})^2(\text{Planck area})$$

$$= - 8\pi(\text{fine structure constant})mc^2/(\text{Compton wavelength})(\text{Planck area})$$

$$\frac{d\Delta_{vac}}{dV} = -\frac{8\pi\alpha mc^2}{\lambda_e L_p^2} = -\frac{8\pi e^2}{\lambda_e^2 L_p^2} \square -10^{73}\frac{Gev}{cm^3} \square -10^{-45}\frac{m_p c^2}{L_p^3}$$

$$\alpha \equiv \frac{e^2}{\hbar c} \square \frac{1}{137}$$

$$\lambda_e \equiv \frac{\hbar}{mc} \square 10^{-11} cm$$

$$mc^2 \square \frac{1}{2}Mev$$

$$L_p^2 \square 10^{-66} cm^2$$

$$(4.1)$$

Note that the ratio $[(\text{Planck area})/(\text{fine structure constant})]^{1/2}$ i.e. $L_p/\sqrt{\alpha}$ is the Kaluza-Klein hyperspace compactification scale. Note that the superconducting vacuum energy gap density whilst enormous by everyday standards is still ultra tiny $\sim 10^{-45}$ relative to the quantum gravity Planck energy density. In my new theory here there is never any direct quantization of the classical gravity field, which is an emergent collective macro-quantum mode out of this spontaneous broken symmetry of the gauge force and source fields in the false unstable micro-quantum vacuum of global special relativity. Thus, I complete what Andre Sakharov started in Moscow in 1967. The theory here can be called the "PV theory of gravity" where "PV" stands for "Polarized Vacuum" – not what Hal Puthoff has called "PV gravity"[345] which is purely a metaphor in a classical phenomenology with no quantum physics in it.

How interesting!

---

[345] http://arxiv.org/abs/gr-qc/9909037

Since everything must be scale-dependent, I make a leap of faith for now that I have not completely justified mathematically, but one can see precognitively[346], i.e., intuitively, where this is all heading. We now see where the "wavelet"[347] macro-quantum order parameter $\Psi(x,p)$ comes from. The phase of $\Psi(x,p)$ gives Einstein's classical Curve World geometrodynamic "wavelet" field $g_{\mu\nu}(x,p)$ along the lines shown by Hagen Kleinert of the Free University of Berlin. The amplitude of $\Psi(x,p)$ give the *renormalized* quintessent "wavelet" $\Lambda^*(x,p)$ field that explains, I say:

1. Why the universe accelerates i.e., $\Lambda^*\left(x, p \to \hbar/10^{28}\,cm\right) > 0$

2. What dark matter is i.e., $\Lambda^*(x,p) < 0$

3. How to make Star Gate time travel machines to past, future and parallel brane worlds like the Universe Next Door[348] of Jacques Vallee's "Magonia" where the UFOs allegedly come from? $\Lambda^*(x,p) > 0$

4. How to make weightless warp drive. (Clue: near field version of Ray Chiao's "gravity radio"[349]?)

---

[346] As in the Puthoff-Targ CIA-funded Remote-Viewing Stanford Research Institute experiments in 1973 when I first met them with Astronaut Edgar Mitchell, Brendan O' Regan and more of the others. See Martin Gardner's "Magic and Paraphysics" in "Science, Good, Bad and Bogus".

[347] There are still a lot of formal details to work out in the replacement of globally flat Fourier integrals by adaptive scale-dependent windowed wavelet transforms required in classical curved spacetime with the breakdown of global flatness of special relativity. So this is a "program" for continued R&D.

[348] A brane parallel universe would have to be less than ~ 2 millimeters away from us in hyperspace according to current measurements. This number is changing with better experiments, e.g. Ch 7 Hawking, Aug 2000 Scientific American "The Universe's Unseen Dimensions".

[349] http://xxx.lanl.gov/abs/gr-qc/0204012.

$$G_{\mu v}^{;v}(x,p) = -\frac{\partial \Lambda^*(x,p)}{\partial x^v} g_{\mu v}(x,p) = T_{\mu v|truevac}^{;v}(x,p) \neq 0$$

$$(4.2)$$

5. Why the arrow of time, i.e. low entropy early universe.

$$S_{truevac}(t=0) = \frac{\rho_n}{\rho_s + \rho_n} S_{falsevac} = L_p^2 \Lambda^* S_{falsevac}$$

$$S_{early-universe}(t=0) \ll S_{older-universe}(t)$$

$$(4.3)$$

This is the real PV (Polarized Vacuum) theory of gravity not the False Idol, that still born thing, that Hal Puthoff has been Hawking.

Don't be put off by skim milk masquerading as cream! ☺[350]

5.   UFO Star Gate Time Travel Warp Drive Physics

Exact supersymmetry is invoked to keep the sum of all zero point energy densities of all fields zero. However supersymmetry is broken so it is not clear if the argument works. The argument would go something like this. Take a toy universe with only electrons and photons. The boson photon has spin 1. Therefore, it has 3 independent spin polarizations tripling the number of field oscillators in a given quantized region of the lattice field oscillator phase space.[351] Each field oscillator has a cell area h in phase space. Each cell holds 2S + 1 field oscillators of a given type where the spin of the field is S. The photon's supersymmetry fermion partner is the spin ½ "photino" with two spin components. There is also the electron field with two more fermion spin components. Finally, the electron has a spin 0 partner the electrino with only one boson spin component.

---

[350] "Things are seldom what they seem." HMS Pinafore, Gilbert and Sullivan.
[351] Given a cell in field oscillator phase space, the cell can either be empty of a real fermion or have one real fermion of a given type in it. That's the Pauli exclusion principle. The same cell can have any number of real bosons of a given type in it. However, the real quanta do not determine micro-quantum vacuum structure until they get very dense.

Therefore, each cell in phase space has four fermion spin components of gravitating negative zero point energy and four boson spin components of anti-gravitating zero point energy. This gives an ordinary $\Lambda = 0$ non-gravitating vacuum. However, if we can control this balance we have something stupendous! This is the End of The Beginning of Our Conquest of Super Cosmos as we transform to "The Masters of Hyperspace".[352]

To be continued.

---

[352] Coined by Michio Kaku in "Hyperspace".

Wigner Scale-Space Density & UFO Technological Surprise

Jack Sarfatti

The Wigner phase space density for a single-particle space for two spacetime events P, P^1 is the following Fourier transform of the reduced single particle quantum density matrix $P^1(P, P^1)$

$$W_1(\mathrm{x}, p) = \frac{1}{(2\pi h)^4} \int e^{ipy/\mathrm{h}} P1\left(x - \frac{y}{2}, x + \frac{y}{2}\right) d^4 y \qquad (1.1)$$

Replace the Fourier transform kernel $e^{ipy/\mathrm{h}}$ by the wavelet-kernel

$$\psi_{s,x}(y) \equiv \frac{1}{|s|^4} \psi\left(\frac{y - x}{s}\right) \psi\left(\frac{y - x}{s}\right) \qquad (1.2)$$

for scale s.

The Wigner scale-space density is then

$$W_1(x, s) = \int \psi_{s,x}(y) P_1\left(x - \frac{y}{2}, x + \frac{y}{2}\right) d^4 y \qquad (1.3)$$

Superfluid ODLRO is factorization of the first reduced density matrix into the macro-quantum phase coherent part with local order parameter $\psi(x) = |\psi(x)| e^{i \arg \psi(x)}$ and residual random noise "normal fluid" fluctuations

345

$$p_1\left(x-\frac{y}{2}, x+\frac{y}{2}\right) \rightarrow \psi *\left(x-\frac{y}{2}\right)\psi\left(x+\frac{y}{2}\right) + p_{1n}\left(x-\frac{y}{2}, x+\frac{y}{2}\right) \quad (1.4)$$

This is generally true for real both real and virtual superfluids.

In the case of my new macro-quantum vacuum physics, the spacetime coordinates are that of the center of mass of a single virtual electron-positron pair[353] in a bound state wave packet $\psi(x)/\sqrt{N_{e+e-BEC}}$ where $N_{e+e-BEC}$ is the total number of virtual electron-positron pair that macroscopically occupy that bound state of lower energy than the random unbound "ionized" virtual pairs of the normal fluid PV QED zero point fluctuations.[354]

The locally variable quintessent field is then

$$\Lambda(x,s) = L_p W_{1ZPF}(x,s) = L_p\left[\frac{1}{L_p^3} - W_{1e+e-BEC}(w,s)\right] \quad (1.5)$$

where

$$W_{1ZPF}(x,s) = \sum W_{1ZPFi}(x,s)$$
$$i = fermions + bosons \quad (1.6)$$

ZPF means locally random "Zero Point Fluctuations" and $L_p^2 \equiv hG/c^3$.

---

[353] The relative coordinate for the separation between the virtual electron and its twin virtual positron in the same bound state pair is integrated out in the usual way like in the BCS and Gorkov Green's function models of real superconductors.

[354] Do not confuse PV zero point fermion virtual pair fluctuations with zero point virtual photon fluctuations. The latter can be viewed as advanced radiation reaction as in Dirac-Wheeler-Feynman model discussed by Hoyle and Narlikar.

$$W_{e+e-BEC}(x,s) = \int \psi_{x,s}(y)\psi * \left(x - \frac{y}{2}\right)\psi\left(x + \frac{y}{2}\right)d^4y \qquad (1.7)$$

$$\Psi^{\%}_{e+e-BEC}(x,s) \equiv \sqrt{\left|W_{e+e-BEC}(x,s)\right|}e^{i\Theta(x,s)} \qquad (1.8)$$

Note that $\psi_{x,s}(y)d^4y$ must be dimensionless – hence my choice of the normalization factor $|s|^{-4}$ for the basic "mother wavelet" in (1.2).

Einstein's "classical"[355] geometrodynamic field is[356], with scale dependence explicit, is

$$g_{\mu\nu}(x,s) = L_p^2\left(\frac{\partial^2\Theta(x,s)}{\partial x^\mu \partial x\nu} + \frac{\partial^2\Theta(x,s)}{\partial x^\nu \partial\mu}\right) \qquad (1.9)$$

Einstein's local field equation is generalized to

$$R_{\mu\nu}(x,s) - \frac{1}{2}R^\sigma_\sigma(x,s)g_{\mu\nu}(x,s) + \Lambda(x,s)g_{\mu\nu}(x,s) = -\frac{8\pi G}{c}T_{\mu\nu}(x,s) \qquad (1.10)$$

supplemented by the covariant Landau-Ginzburg equation with covariant derivatives D_μ

$$D^\mu D_\mu \Psi^{\%}_{e+e-BEC}(x,s) + \alpha \Psi^{\%}_{e+e-BEC}(x,s) + \beta\left|\Psi^{\%}_{e+e-BEC}(x,s)\right|^2$$

$$\Psi^{\%}_{e+e-BEC}(x,s) = 0$$

$$(1.11)$$

---

[355] Really "macro quantum", i.e., strictly speaking there is no "classical limit" in old sense of Bohr's Correspondence Principle.
[356] Neglecting gauge force field contributions now for simplicity of the essential new idea here.

The usual Bianchi identities leading to

$$D^\nu G_{\mu\nu} \equiv D^\nu \left(R_{\mu\nu} - \frac{1}{2} R^\sigma_\sigma g_{\mu\nu} \right) = 0 \qquad (1.12)$$

are violated when the quintessent field Λ[357] is locally variable. These Bianchi identities associated with local conservation of momenergy for the external field stress energy density tensor, i.e.,

$$D^\nu T_{\mu\nu} = 0 \qquad (1.13)$$

These Bianchi identities assume two physical conditions

1. metricity, i.e. $D^\nu g_{\mu\nu} = 0$

2. zero torsion tensor, i.e. $\Gamma^\lambda_{\mu\nu} = \Gamma^\lambda_{\nu\mu}$

Since the spacetime stiffness factor G/c^4 is so huge, and since Λ seems to[358] depend on the nonradiating electromagnetic induction near fields in a softer more manageable environmentally friendly way, we make the approximation to Einstein's generalized field equation

---

[357] The usual macro-quantum vacuum has zero quintessence. Positive quintessence with overpowering negative quantum pressure consequently *anti-gravitates* as exotic vacuum stuff. Therefore, this simply and elegantly explains several important, so far unexplained, phenomena: acceleration of universe; star gate time travel; and the weightless warp drive of the alleged flying saucers. The latter alleged phenomenon, if real, is a major military threat from an advanced non-terrestrial culture that the USAF has no defense against at this time. Negative quintessence with overpowering positive quantum pressure consequently gravitates, thus explaining the dark energy of the "missing mass" of the universe, which is most of the mass of the universe.
[358] Conjecture at this stage deduced from alleged flying saucer flight capability.

$$G_{\mu\nu}(x,s) + \Lambda(x,s)g_{\mu\nu}(x,s); 0 \qquad (1.14)$$

This leapfrogs over the space-time stiffness barrier completely. In this limit

$$D^\nu G_{\mu\nu}(x,s) + \frac{\partial \Lambda(x,s)}{\partial x^\nu} g_{\mu\nu}(x,s); 0 \qquad (1.15)$$

Equation (1.15) is the "vacuum propeller" equation for weightless warp drive.[359] Here I assume metricity. Therefore, in this case the locally variable quintessent field is, perhaps a torsion field generator.[360] That is, make the wild "half-baked" conjecture

$$\frac{\partial \Lambda}{\partial x^\mu} \overset{?}{=} T^\nu_{\mu\nu} \qquad (1.16)$$

Where $T^\lambda_{\mu\nu}$ is the third rank torsion field tensor corresponding to dislocation gap defects in Hagen Kleinert's "World Crystal Lattice" elasticity strain model of Einstein's 1915 Geometrodynamics as in equation (1.9) above. The infinitesimal curvature loops of parallel transport of tangent vector fields fail to close.

If we keep the stress energy density tensor, we have the condition of either extracting vacuum energy for traditional power uses including impulse propulsion of the usual rocket kind, and also concealing energy storing it inside the macro-quantum vacuum for stealth cloaking of military air force and naval operations including silent running of submarines.

---

[359] The flying saucer generates its own timelike geodesic for the motion of its center of mass with small tidal forces and no local g-forces.
[360] Gennady Shipov in Moscow has been working on such a theory.

Index of names of people and key words.*

charlatans, *129*
Charles Bennett, *147*
Charles Darwin, 207
Charles Lindbergh, 37
Charles Misner, 91, 92
Charlie Chaplin, *71*
cheerful facts, 222
Cherubim, *95*
Cheryl Haley, 114
Cheryl Haley and Richard Feynman at Esalen, Big Sur, CA, 115
Cheryl Haley with Jack Sarfatti at the Cal Tech Kip Fest 2000, 116
Chestnut Street, 20
Chicago Options Market, **44**
Chickering Letter, 147
Child Buyer, 39
childhood ecstasy, 119
child-like faith in God, 239
Chinese, 48
Chris Bird, 55
Chris Nunn, 213
Christ, 245, 247
Christ's Home, 119
Christian Chivalry, 7
Christmas, 1941, 13
chronology protection conjecture, 18, 294
CIA, 15, 58, 130, 173, 341
CIA Chief of Station, 15, **51**, 57
CIA Los Angeles Office, 57
CIA Memorandum for the Record, *147*
CIA Memorandum for the Record on Jack's ideas, 137
CIA Science Technology, *141*
CIA's Kit Green, 110
Ciao Manhattan, *77*, 79
Cipher of Genesis, *5, 20, 55*
citadel, 260
CITY, *105*
City Light's Bookstore, *64*
Civil Air Patrol, *40*, 41

Civil War, *71*, 79
Civil War General, William Heine, **71**
civil-rights sit-in, 119
claiming to be ET, 101
Clara Zetkin, *66*
Clarke, *164*
classical, 34, 78, 138, 157
classical action, 324
classical bit, 139
classical curved spacetime, 201, 205, 265, 270, 274, 276, 288
classical electromagnetic field, 282
classical field, 321
Classical field theories, 6
classical mechanical action, 327
classical probabilities, 18
classical vacuum, 34
Claudio Naranjo, 110
Clinton, *164*
Clive Prince, *129*
cloak effect, *35*
clone a photon, 320
Close Encounters, 137
Close Encounters of the Third Kind, *91*
closed time-like curves, *95*
closely packed, 339
closest cronies, **71**
CNN, 15
cobalt, 120
Cocteau, *51*
coherence, 83, 265, 276, 293
coherence order parameter, 205
coherent order, 274, 334
coherent phase, *47*
coherent phase ordered quantum vacuum, *60*
coherent quantum superpositions, *60*
coherent virtual superfluid, 82
Col. Bryant E. Moore, 13
Col. John Alexander, **57**

Feynman's insight, 211
Feynman's Ph.D, 209
fiber bundle, 68
field functions of spacetime, 6
field point, 320
Fifth Avenue, *71*, 245
fighter jet, *35*
final cause, 208
final causes, 206, 207
Final Secret of the Illuminati, 174
final states, 44
Finger of Fate, 213
finite loss of fidelity, 219
fire, 83
fire escape, 30
First Crusade, *53*
First Earth Battalion, **53**
first man, 87
First Wave, 90, 140
Flap Doodle, 147
Flat Earth Physics, 205
Flat World, 269, 271, 272, 274, 277,
    279, 281, 283, 332, 333, 338
Flatbush, *24, 162*
Flatbush, Brooklyn, *94*
Flatlanders, 23
Fledermaus, 170
Floyd Bennett Naval Air Station, 40
flux lines, 83
Flying Castle of the ET Giant, 11
flying saucer, 4, *20, 33, 35*, 61, 62,
    190, 298, 326
Flying Saucer Physics, 33
flying saucers, 18, 195
Flying Saucers, 60
flywheels from Lawrence Livermore
    Laboratory, 147
foam, 35
Fock space, 219
force, 34, 80, 84, 89, 157
forced quantum harmonic oscillator,
    209

Ford Philco Aeronutronics, 331
forgeries, 33
Forza del Destino, 213
foundations of physics, 42
Founding Fathers, 208
Fountainhead, 37, 132
four hundred, 27
Four Noble Truths, 241
Four Questions, 90
Fourier integrals, 341
fractional electric charge, *60*
fragmentary view of the whole, 191
France, **50, 53**
Francis Ford Coppola, *40, 48*, **52**, *72*,
    90, 105, 111, 120, *140*
Francois Truffaut, 91
Frank Barr, 113
Frank Lauria, i
Frank Sinatra, 60
Franklin D. Roosevelt, *71*
Franz Borkenau, 76
fraud, 98
freak auto accident, 48
Fred Alan Wolf, *20*, 43, 44, 46, 47,
    48, *53*, 54, 91, *99, 105*, 109, 110,
    201
Fred Hoyle, 81
Fred W. Cummings, 36, 331, 336
free float, 35, *48, 62*
free float frames, *34*
free float geodesic, 92
free float timelike geodesic, 265
Free University of Berlin, 98, 341
free will, 19, 207
Freemasons, *130*
French Intelligence, 52
French Military Intelligence, 109
French Naval Attaché in Rome, **48**
French Navy, *48*
French physicist, *109*
French Rabbi, 63
French Rationalism, 200

Newtonian absolute simultaneity, 282
Newtonian classical mechanics, 111
Newtonian limit, 322
Newtonian mechanics, 112
nice Jewish boy, 244
Nick Cook, **62**
Nick Herbert, *20*, *95*, *105*, *106*, *107*, *110*, 111, *139*, 326
Niels Bohr, 18
Nietzsche, 24, *88*
Nirvana, 241
no joke, 28
no shock waves, 33
Nobel Prize, *45*, *60*
no-cloning theorem, 147, 217
No-Cloning Theorem, 321
Noetic Sciences, *20*
nonclassical phase, 275
non-equilibrium, 202
nonlethal weapons, 190
nonlinear and nonlocal potential terms, 286
nonlocal, *114*
nonlocal quantum bit potential, 108
nonlocal quantum connections, 23
nonlocally connected, *60*
nonlocally entangled state, 279
nonradiating black holes, 44
nonrandom, 274, 279
nonrandom correlations, 205
nontraversable wormholes, *92*
normal fluid, 329, 334, 335
normal metal, 23
Norman Quebedeau, 165
Norman Thomas, 147
North Beach, 4, 7, *20*, 45, *63*, *69*, *72*, *93*, *109*, 113, 121, *131*, *134*, *141*, *161*
North Beach cafes, 2
North Beach Magazine, *64*, 65
Norwood Pratt, **69**

Novecento, *66*
novel properties and concepts, 83
Novosibirsk, *91*
Noyes Lodge, 79, 161
NSF, 194
NSF Summer Institute, **43**
nuclear attack, 138
nuclear explosion, 36
nuclear long-lived isomer, 36
nuclear weapons, *37*, *140*
nuclear winter, 147
nucleon, 60
numbers, physics, and consciousness, 118
Numismatist, 36
Nuovo Cimento, 330
nut desk, 112
Nutty Universe, 326
NYPD homicide, 32
O Brane New World, 19, 95, 205
O Brane World, 216
O. J. Simpson, *55*
O'Regan and Breen, **47**
obelisk, *77*
Oberlin Conservatory, 170
objective and local, *107*
objective science, 88
objectively real, 322
observer independent, *107*
Occidental College, 18, **50**
occult, **50**, **52**, **70**, **130**, **163**, **164**
Occult SS, *50*, 52
Occult Third Reich, *69*, *162*
October Sky, *36*, *38*
October Surprise, 113
Odd Couple, **44**
Odeon, *54*
ODLRO, 316
off mass shell, 204
off-diagonal time loops, 214
Office of Scientific Research and Development, 37

Ramadan, 240, 241
RAND, 83
random, 35
random electromagnetic zero point
    fluctuations, 82
random zero point fluctuations, 23
rank, 272
rape, 238
Rasa Gustaitus, *105*
Rashi, *52, 53*
Rashi de Troyes, *53, 63, 64, 68*
Rashi des Troyes, *54*
Rashi's daughters, 63
Rat Pack, 45
Ray Chiao, 147, 341
Ray Stanford, *102*
RCA's Spectra, *100*
reach the stars, *129*
Reagan Administration, 142
Reagan White House, 113
real and not quantum virtual, 92
real entangled electron pairs, *60*
real number axis, 202
real on mass shell, 223
reality, 34, 86, 89, 164, 239, 246,
    253, 256, 257, 259
Reason, 260
receptive minds, 27
recklessness, 89
reconciler, *89*
red and blue shifts from UFOs, 325
red laser, *103*
red shifted, 325
redshift functions, 293
reducto ad absurdum, 218
regenerate heart tissue, 12
Regge trajectories, 82
Regina, 239
Regina Sarfatti, 66, 238
reincarnate, 259
reincarnation, *55, 69, 88, 102, 174,*
    241

relativism, *69*
relay switching circuits, 24
religion, *88, 129, 141*
Remote Viewers, 102
Remote Viewers: The Secret History of
    America's Psychic Spies, 173
remote viewing, 15, *20, 48, 53, 103,*
    139
Remote Viewing, 15
Remote-Viewing, 341
remote-viewing research, 109
Rene Thom's catastrophe theory,
    111
renormalization, 267, 268, 294
renormalization factor, 332, 333
renormalization group, 83
renormalization trick, *60*
Republican Party, *130*
Resource Letter on Symmetry in
    Physics, 81, 336
Restoration of Israel, 70
restore the broken symmetry, 83
retarded, 213
retro PK, 217
retroactive superluminal
    communication, *44*
retroactivity, 202
retrocausality, 198
Rev. Moon, *68*
Revelle College, *60*
reversible, 218
Review of Modern Physics, 44
revolutionary discoveries, 42
Revolutionary War, 238
rich American in the Land Rover,
    64
Richard DeLauer, 147
Richard Dolan, 91
Richard Farina, *160*
Richard Feynman, 18, *57, 61,* 92,
    114, 209, 266
Richard Gott, 173

385

* You may have to look for first name, e.g. "Brendan O'Regan", rather than last name "O'Regan".

About the Author

The reading and movie-going public's interest in eccentric rebel physics and math geniuses seen in "A Beautiful Mind" about John Nash, and Stephen Wolfram's "A New Kind of Science" is keen these days. A dashing Gilbert and Sullivan tenor in his youth, urbane, yet Jewish from The Mean Streets of Flatbush, Ivy League Cornell-educated Jack Sarfatti with a physics Ph.D. from the University of California certainly qualifies. His true-life "cloak and dagger, skull and bones" "UFO Black Ops" cliff-hanging adventure story with est's Werner Erhard, Ira Einhorn, Tim Leary, Esalen, Reagan "Star Wars" Think Tank "Buttoned Down Bohemians", Francis Ford Coppola, Hollywood movie billionaire Marshall Naify and many beautiful intelligent women, is more like James Bond's than Stephen Hawking's. Jack's unique truly original real physics ideas on time travel, metric engineering of Star Gates to the parallel universes next door in "Super Cosmos", weightless warp drives making Star Trek real, contact with alien intelligence and consciousness, take Hawking's ideas to the next level. Lavishly illustrated with photographs, corroborating historical documents and original cartoons.

"With infectious enthusiasm for his subject, Jack Sarfatti explains how physics has replaced philosophy as an over-arching discipline that spans the once discontinuous worlds of science and the humanities. Since he first came to the Bay Area in the mid-1970s, physicist Jack Sarfatti has been a provocative presence in local intellectual life.

Leaping from North Beach cafes to leading policy think tanks, he has cut a broad intellectual swath, challenging the preconceptions of poets, political thinkers and physicists alike.

With a background in quantum theory, he claims to break new ground in scientific understanding of the eternal questions: "Who are we? Where do we come from? Where are we going?"

In this interview, he discusses the breakdown of a paradigm that, for centuries in the West, has viewed science and humanistic thought as irrevocably separated." The Universe, As Seen From North Beach by Stephen Schwartz, *San Francisco Chronicle*, August 17, 1997

This book is, by far, the weirdest true story ever told. It tells how our Super Cosmos, of parallel universes floating in hyperspace, is connected together by Star Gate time travel machines carrying "The Masters of Hyperspace"[1] back to us from our futures. We all come into being and becoming from the post-quantum Mind of God. Participate in the real life thrilling Psi Wars adventure, the amazing paranormal flying saucer Grail Quest of visionary physicist Jack

Sarfatti and "the others" as they decode The Cabala's[2] "Cipher of Genesis" demystifying the enigma of consciousness, a body guard of spiritual truths wrapped in the secrets of the material world of post-quantum physics.

Jack Sarfatti was born in Brooklyn New York, on September 14, 1939. He graduated Woody Allen's Alma Mater, Midwood High School in 1956. Sent to Cornell on full scholarship, a classmate of Tom Pynchon's (*Gravity's Rainbow, V, Vineland*), Jack studied with the men who built the atomic bomb at Los Alamos graduating with a BA in physics in 1960. Jack went on to get a Ph.D. in physics from the University of California. Jack was a physics professor at San Diego State University where he met and worked with Fred Alan Wolf, winner of American Book Award 1982 for *Taking The Quantum Leap*. Jack and Fred co-wrote *Space-Time and Beyond* (Dutton, 1975) which sold well. The literary agent for that book was the notorious Ira Einhorn. Jack, as leader of the est-Esalen Institute sponsored Physics Consciousness Research Group in 1976, brought his roommate Gary Zukav there, taught him a smattering of physics, wrote and closely edited most of the real physics content of *The Dancing Wu Li Masters* including rewrites of material contributed by Henry Stapp, David Finkelstein and other physicists. Jack has been President of the Internet Science Education Project in San Francisco since 1995 *"In the half-century that North Beach has supplied a haven for waves of brilliant and erratic thinkers, few have rivaled Jack Sarfatti for comprehensive weirdness and what just might be original genius"*—Jerry Carroll in "Another Eccentric Genius in North Beach?" *San Francisco Chronicle*, May 11,1981, p.23

[1]Michio Kaku's book *Hyperspace* and Stephen Hawking's *The Universe in a Nutshell*.

[2] Hidden knowledge common to Jewish, Christian and Islamic Mystics, see Erik Davis's *Techgnosis* and the books of Joseph Campbell on Myths.

Printed in November 2023
by Rotomail Italia S.p.A., Vignate (MI) - Italy